CliffsQuickReview™
Biology

By I. Edward Alcamo, PhD and
Kelly Schweitzer, PhD

Wiley Publishing, Inc.

About the Authors

I. Edward Alcamo, PhD is a Distinguished Professor of Microbiology at the State University of New York at Farmingdale. He has taught at the college level for 30 years, specializing in microbiology for nursing and other allied health science students.

Kelly Schweitzer received her PhD from Indiana University School of Medicine in Biochemistry and Molecular Biology. She is currently a postdoctoral fellow at Indiana University School of Medicine in the Pathology Dept. Her research focuses on the molecular biology of Alzheimer's Disease.

Publisher's Acknowledgments

Editorial

Senior Project Editor: Joan Friedman

Acquisitions Editor: Sherry Gomoll

Copy Editor: Esmeralda St. Clair

Editorial Assistant: Jennifer Young

Production

Indexer: TECHBOOKS Production Services

Proofreader: Sossity R. Smith

Wiley Indianapolis Composition Services

CliffsQuickReview™ *Biology*

Published by:
Wiley Publishing, Inc.
909 Third Avenue
New York, NY 10022
www.wiley.com

Copyright © 2001 Wiley Publishing, Inc., New York, New York
Library of Congress Control Number: 2001016953
ISBN: 0-7645-6375-0
Printed in the United States of America
10 9 8 7 6
1O/QR/RR/QS/IN
Published by Wiley Publishing, Inc., New York, NY
Published simultaneously in Canada

Table of Contents

INTRODUCTION

This book provides a concise, comprehensive review of the key concepts of the study of living things. Topics covered in detail include the atoms and molecules that comprise all living organisms; the central dogma of biology; the three central molecules (DNA, RNA, and protein); and the most vital biological processes, such as mitosis, meiosis, and reproduction. Subdisciplines of biology are also thoroughly explained, including taxonomy, anatomy, and ecology. A review of genetics and evolution includes the controversial and groundbreaking theories of Gregor Mendel and Charles Darwin. The biochemical processes of respiration and photosynthesis are discussed in detail for comparison, and several chapters focus on plant structure and function. Due to increasing interest in recombinant biotechnology, Chapter 11 highlights common molecular biology techniques used to identify genetic disorders and combat disease in humans. Finally, this book provides several chapters reviewing comparative anatomy of humans and other organisms.

CliffsQuickReview Biology includes bare bones information about a wide variety of biological processes, from microbiology and biochemistry to evolution and ecology.

Why You Need This Book

Can you answer yes to any of these questions?

- Do you need to review the fundamentals of biology fast?
- Do you need a course supplement to aid your study of biology?
- Do you need a concise, comprehensive reference for your biology studies?

If so, then *CliffsQuickReview Biology* is for you!

How to Use This Book

You can use this book in any way that fits your personal style for study and review; you decide what works best with your needs. You can either read the book from cover to cover or just look for the information you want and put it back on the shelf for later. Here are just a few ways you can search for topics:

- Use the Pocket Guide to find essential information, such as formulas for respiration and photosynthesis and the structures of general amino acids.

- Look for areas of interest in the book's Table of Contents, or use the index to find specific topics.

- Flip through the book, looking for subject areas at the top of each page.

- Get a glimpse of what you'll gain from a chapter by reading through the "Chapter Check-In" at the beginning of each chapter.

- Use the Chapter Checkout at the end of each chapter to gauge your grasp of the important information you need to know.

- Test your knowledge more completely in the CQR Review and look for additional sources of information in the CQR Resource Center. The Resource Center provides additional biology references that you can use to build on the concepts reviewed in this book.

- Use the glossary to find key terms fast. This book defines new terms and concepts where they first appear in the chapter. If a word is bold-faced, you can find its definition in the book's glossary.

- Or flip through the book until you find what you're looking for. We organized this book to gradually build on key concepts.

Visit Our Web Site

A great resource, www.hungryminds.com features review materials, valuable Internet links, quizzes, and more to enhance your learning. The site also features timely articles and tips, plus downloadable versions of many CliffsNotes books.

When you stop by our site, don't hesitate to share your thoughts about this book or any Hungry Minds product. Just click the Talk to Us button. We welcome your feedback!

Chapter 1

THE SCIENCE OF BIOLOGY

Chapter Check-In

❑ Identifying the characteristics of living things

❑ Mapping out the scientific method

Biology is the study of living things. It encompasses the cellular basis of living things, the energy metabolism that underlies the activities of life, and the genetic bases for inheritance in organisms. Biology also includes the study of evolutionary relationships among organisms and the diversity of life on earth. It considers the biology of microorganisms, plants, and animals, and it brings together the structural and functional relationships that underlie their day-to-day activities. Biology draws on the sciences of chemistry and physics for its foundations and applies the laws of these disciplines to living things.

Many subdisciplines and special areas of biology exist, which can be conveniently divided into practical and theoretical categories. Types of *practical biology* include plant breeding, wildlife management, medical science, and crop production. *Theoretical biology* encompasses such disciplines as *physiology* (the study of the function of living things), *biochemistry* (the study of the chemistry of organisms), *taxonomy* (classification), *ecology* (the study of populations and their environments), and *microbiology* (the study of microscopic organisms).

Characteristics of Living Things

Defining a living thing is a difficult proposition, as is defining "life"—that property possessed by living things. However, a living thing possesses certain properties that help define what life is.

Complex organization

Living things have a level of complexity and organization not found in lifeless objects. At its most fundamental level, a living thing is composed of one or more **cells.** These units, generally too small to be seen with the unaided eye, are organized into tissues. A *tissue* is a series of cells that accomplish a shared function. Tissues, in turn, form *organs,* such as the stomach and kidney. A number of organs working together compose an *organ system.* An organism is a complex series of various organ systems.

Metabolism

Living things exhibit a rapid turnover of chemical materials, which is referred to as **metabolism.** Metabolism involves exchanges of chemical matter with the external environment and extensive transformations of organic matter within the cells of a living organism. Metabolism generally involves the release or use of chemical energy. Nonliving things do not display metabolism.

Responsiveness

All living things are able to respond to stimuli in the external environment. For example, living things respond to changes in light, heat, sound, and chemical and mechanical contact. To detect stimuli, organisms have means for receiving information, such as eyes, ears, and taste buds.

To respond effectively to changes in the environment, an organism must coordinate its responses. A system of nerves and a number of chemical regulators called **hormones** coordinate activities within an organism. The organism responds to the stimuli by means of a number of effectors, such as muscles and glands. Energy is generally used in the process.

Organisms change their behavior in response to changes in the surrounding environment. For example, an organism may move in response to its environment. Responses such as this occur in definite patterns and make up the behavior of an organism. The behavior is active, not passive; an animal responding to a stimulus is different from a stone rolling down a hill. Living things display responsiveness; nonliving things do not.

Growth

Growth requires an organism to take in material from the environment and organize the material into its own structures. To accomplish growth, an organism expends some of the energy it acquires during metabolism. An organism has a pattern for accomplishing the building of growth structures.

During growth, a living organism transforms material that is unlike itself into materials that are like it. A person, for example, digests a meal of meat and vegetables and transforms the chemical material into more of himself or herself. A nonliving organism does not display this characteristic.

Reproduction

A living thing has the ability to produce copies of itself by the process known as *reproduction.* These copies are made while the organism is still living. Among plants and simple animals, reproduction is often an extension of the growth process. For example, bacteria grow and quickly reach maturity, after which they split into two organisms by the process of *asexual reproduction.* Asexual reproduction involves only one parent, and the resulting cells are generally identical to the parent cell.

More complex organisms engage in a type of reproduction called *sexual reproduction,* in which two parents contribute to the formation of a new individual. During this process, a new combination of traits can be produced. The process is generally more complex than asexual reproduction, requiring that parents find one another, then (usually) care for their offspring before it can live independently. Nonliving things have no such ability or requirements.

Evolution

Populations of living organisms have the ability to adapt to their environment through the process of evolution. During evolution, changes occur in populations, and the organisms in the population become better able to metabolize, respond, and reproduce. They develop abilities to cope with their environment that their ancestors did not have.

Evolution also results in a greater variety of organisms than existed in previous eras. This proliferation of populations of organisms is unique to living things.

Ecology

The environment influences the living things that it surrounds. *Ecology* is the study of relationships between organisms and their relationships with their environment. Living things can alter their environment, but nonliving things cannot. Living things, for example, may migrate or hibernate if the environment becomes difficult to live in.

Scientific Method

Biology is one of the major sciences. Scientists have acquired biological knowledge through processes known as **scientific methods.** There is no one scientific method. The steps of a scientific method make up an orderly way of gaining information about the biological world. The knowledge gained is sometimes useful in solving particular problems and is sometimes simply of interest without any practical application at the time.

A scientific method requires a systematic search for information by observation and experimentation. The basic steps of any scientific method are stating a problem, collecting information, forming a hypothesis, experimenting to test the hypothesis, recording and analyzing data, and forming a conclusion.

Observation

The first step in a scientific method is stating a problem based on observation. In this stage, the scientist recognizes that something has happened and that it occurs repeatedly. Therefore, the scientist formulates a question or states a problem for investigation. The next step in the scientific method is to explore resources that may have information about that question or problem. Here, the scientist conducts library research and interacts with other scientists to develop knowledge about the question at hand.

Hypothesis, experimentation, and analysis

Next, a hypothesis is formed, meaning that the scientist proposes a possible solution to the question, realizing that the answer could be incorrect. The scientist tests the hypothesis through experiments that include experimental and control groups. Data from the experiments is collected, recorded, and analyzed.

Conclusion

After analyzing the data, the scientist draws a conclusion. A valid conclusion must be based on the facts observed in the experiments. If the data from repeated experiments support the hypothesis, the scientist will publish the hypothesis and experimental data for other scientists to review and discuss.

Theory and law

Other scientists may not only repeat the experiments but may carry out additional experiments to challenge the findings. If the hypothesis is tested and confirmed often enough, the scientific community calls the hypothesis a **theory.** Then numerous additional experiments test the theory using rigorous experimental methods. Repeated challenges to the theory are presented. If the results continue to support the theory, the theory gains the status of a scientific law. A scientific law is a uniform or constant fact of nature. An example of a law of biology is that all living things are composed of cells.

Chapter Checkout

Q&A

1. List and define the steps of the scientific method.

2. Describe the difference between a theory and a law.

Chapter 2

THE CHEMICAL BASIS OF LIFE

Chapter Check-In

❑ Understanding basic chemical principles

❑ Comparing organic compounds

or many centuries, biology was the study of the natural world. Biologists searched for unidentified plants and animals, classified them, and studied their anatomy and how they acted in nature. Then in the 1700s, scientists discovered the chemical and physical basis of living things. They soon realized that the chemical organization of all living things is remarkably similar.

Elements

All living things on earth are composed of fundamental building blocks of matter called **elements.** More than 100 elements are known to exist, including those that are man-made. An element is a substance that cannot be chemically decomposed. Oxygen, iron, calcium, sodium, hydrogen, carbon, and nitrogen are examples of elements.

Atoms

Each element is composed of one particular kind of atom. An **atom** is the smallest part of an element that can enter into combinations with atoms of other elements.

Atoms consist of positively charged particles called *protons* surrounded by negatively charged particles called *electrons*. A third type of particle, a *neutron,* has no electrical charge; it has the same weight as a proton. Protons and neutrons adhere tightly to form the dense, positively charged nucleus of the atom. Electrons spin around the nucleus.

The electron arrangement in an atom plays an essential role in the chemistry of the atom. Atoms are most stable when their outer shell of electrons has a full quota. The first electron shell has a maximum of two electrons. The second and all other outer shells have a maximum of eight electrons. Atoms tend to gain or lose electrons until their outer shells have a stable arrangement. The gaining or losing of electrons, or the sharing of electrons, contributes to the chemical reactions in which an atom participates.

Molecules

Most of the compounds of interest to biologists are composed of units called molecules. A **molecule** is a precise arrangement of atoms, and a **compound** is a collection of molecules. A molecule may be composed of two or more atoms of the same element, as in oxygen gas, O_2, or it may be composed of atoms from different elements. The arrangements of the atoms in a molecule account for the properties of a compound. The molecular weight is equal to the atomic weights of the atoms in the molecule.

The atoms in molecules may be joined to one another by various linkages called *bonds.* One example of a bond is an *ionic bond,* which is formed when the electrons of one atom transfer to a second atom. This creates electrically charged atoms called *ions.* The electrical charges cause the ions to be attracted to one another, and the attraction forms the ionic bond.

A second type of linkage is a *covalent bond.* A covalent bond forms when two atoms share one or more electrons with one another. For example, as shown in Figure 2-1, oxygen shares its electrons with two hydrogen atoms, and the resulting molecule is water (H_2O). Nitrogen shares its electrons with three hydrogen atoms, and the resulting molecule is ammonia (NH_3). If one pair of electrons is shared, the bond is a single bond; if two pairs are shared, then it is a double bond.

Figure 2-1 Formation of a covalent bond in water and ammonia molecules. In each molecule, the second shell fills with eight electrons.

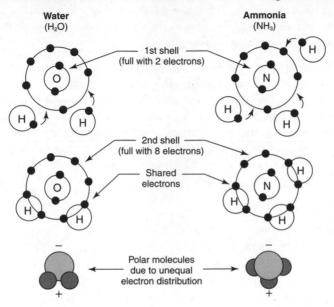

Acids and Bases

Acids are chemical compounds that release hydrogen ions (H+) when placed in water. For example, when hydrogen chloride is placed in water, it releases its hydrogen ions and the solution becomes hydrochloric acid.

Bases are chemical compounds that attract hydrogen atoms when they are placed in water. An example of a base is sodium hydroxide (NaOH). When this substance is placed in water, it attracts hydrogen ions, and a basic (or alkaline) solution results as hydroxyl (—OH) ions accumulate.

Organic Compounds

The chemical compounds of living things are known as *organic compounds* because of their association with organisms. Organic compounds, which are the compounds associated with life processes, are the subject matter of organic chemistry. Among the numerous types of organic compounds, four major categories are found in all living things: carbohydrates, lipids, protein, and nucleic acids.

Carbohydrates

Almost all organisms use carbohydrates as sources of energy. In addition, some carbohydrates serve as structural materials. Carbohydrates are molecules composed of carbon, hydrogen, and oxygen; the ratio of hydrogen atoms to oxygen atoms is 2:1.

Simple carbohydrates, commonly referred to as *sugars,* can be **monosaccharides** if they are composed of single molecules, or **disaccharides** if they are composed of two molecules. The most important monosaccharide is **glucose,** a carbohydrate with the molecular formula $C_6H_{12}O_6$. Glucose is the basic form of fuel in living things. It is soluble and is transported by body fluids to all cells, where it is metabolized to release its energy. Glucose is the starting material for cellular respiration, and it is the main product of photosynthesis (see Chapters 5 and 6).

Three important disaccharides are also found in living things: maltose, sucrose, and lactose. Maltose is a combination of two glucose units covalently linked. The table sugar sucrose is formed by linking glucose to another monosaccharide called *fructose.* (Figure 2-2 shows that in the synthesis of sucrose, a water molecule is produced. The process is therefore called a *dehydration.* The reversal of the process is *hydrolysis,* a process in which the molecule is split and the elements of water are added.) Lactose is composed of glucose and galactose units.

Figure 2-2 Glucose and fructose molecules combine to form the disaccharide sucrose.

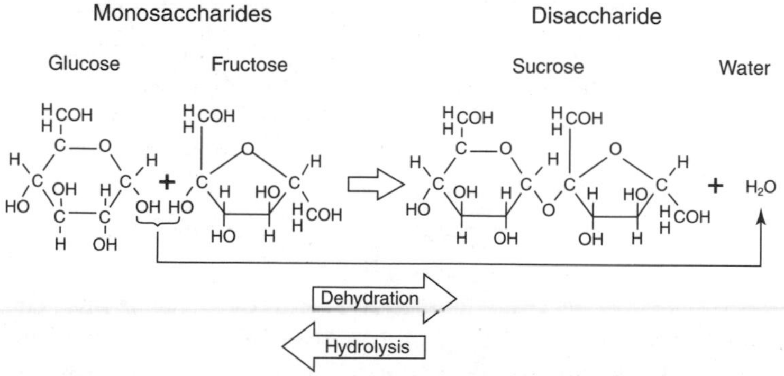

Complex carbohydrates are known as **polysaccharides.** Polysaccharides are formed by linking innumerable monosaccharides. Among the most important polysaccharides are the starches, which are composed of hundreds or thousands of glucose units linked to one another. Starches serve as a storage form for carbohydrates. Much of the world's human population satisfies its energy needs with the starches of rice, wheat, corn, and potatoes.

Two other important polysaccharides are glycogen and cellulose. **Glycogen** is also composed of thousands of glucose units, but the units are bonded in a different pattern than in starches. Glycogen is the form in which glucose is stored in the human liver. Cellulose is used primarily as a structural carbohydrate. It is also composed of glucose units, but the units cannot be released from one another except by a few species of organisms. Wood is composed chiefly of cellulose, as are plant cell walls. Cotton fabric and paper are commercial cellulose products.

Lipids

Lipids are organic molecules composed of carbon, hydrogen, and oxygen atoms. The ratio of hydrogen atoms to oxygen atoms is much higher in lipids than in carbohydrates. Lipids include steroids (the material of which many hormones are composed), waxes, and **fats.**

Fat molecules are composed of a glycerol molecule and one, two, or three molecules of fatty acids (see Figure 2-3). A glycerol molecule contains three hydroxyl (—OH) groups. A **fatty acid** is a long chain of carbon atoms (from 4 to 24) with a carboxyl (—COOH) group at one end. The fatty acids in a fat may be all alike or they may all be different. They are bound to the glycerol molecule by a process that involves the removal of water.

Certain fatty acids have one or more double bonds in their molecules. Fats that include these molecules are *unsaturated fats.* Other fatty acids have no double bonds. Fats that include these fatty acids are *saturated fats.* In most human health situations, the consumption of unsaturated fats is preferred to the consumption of saturated fats.

Fats stored in cells usually form clear oil droplets called *globules* because fats do not dissolve in water. Plants often store fats in their seeds, and animals store fats in large, clear globules in the cells of adipose tissue. The fats in adipose tissue contain much concentrated energy. Hence, they serve as a reserve energy supply to the organism. The enzyme lipase breaks down fats into fatty acids and glycerol in the human digestive system.

Figure 2-3   A fat molecule is constructed by combining a glycerol molecule with three fatty acid molecules. (Two saturated and one unsaturated fatty acids are shown for comparison.) The constructed molecule is at the bottom.

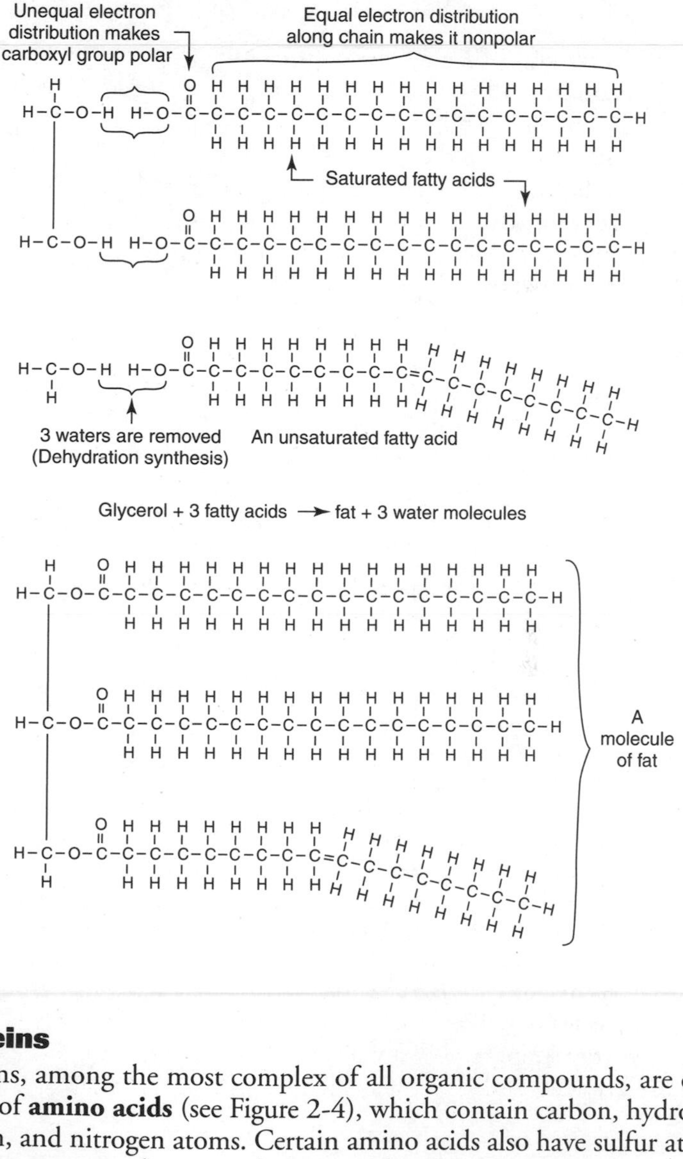

Unequal electron distribution makes carboxyl group polar

Equal electron distribution along chain makes it nonpolar

Saturated fatty acids

3 waters are removed (Dehydration synthesis)

An unsaturated fatty acid

Glycerol + 3 fatty acids ⟶ fat + 3 water molecules

A molecule of fat

Proteins

Proteins, among the most complex of all organic compounds, are composed of **amino acids** (see Figure 2-4), which contain carbon, hydrogen, oxygen, and nitrogen atoms. Certain amino acids also have sulfur atoms, phosphorous, or other trace elements such as iron or copper.

Figure 2-4   The structure and chemistry of amino acids. When two amino acids are joined in a dipeptide, the –OH of one amino acid is removed, and the –H of the second is removed. A dipeptide bond (right) forms to join the amino acids together.

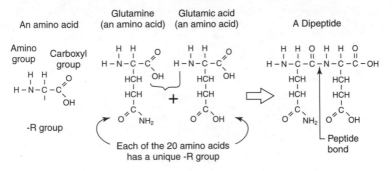

Many proteins are immense in size and extremely complex. However, all proteins are composed of long chains of relatively simple amino acids. There are 20 kinds of amino acids. Each amino acid (see the left illustration in Figure 2-4) has an amino (—NH₂) group, a carboxyl (—COOH) group, and a group of atoms called an —R group (where R stands for *radical*). The amino acids differ depending on the nature of the —R group, as shown in the middle illustration of Figure 2-4. Examples of amino acids are alanine, valine, glutamic acid, tryptophan, tyrosine, and histidine.

The removal of water molecules links amino acids to form a protein. The process is called *dehydration synthesis,* and a byproduct of the synthesis is water. The links forged between the amino acids are *peptide bonds,* and small proteins are often called **peptides.**

All living things depend on proteins for their existence. Proteins are the major molecules from which living things are constructed. Certain proteins are dissolved or suspended in the watery substance of the cells, while others are incorporated into various structures of the cells. Proteins are also found as supporting and strengthening materials in tissues outside of cells. Bone, cartilage, tendons, and ligaments are all composed of protein.

One essential use of proteins is in the construction of enzymes. **Enzymes** catalyze the chemical reactions that take place within cells. They are not used up in a reaction; rather, they remain available to catalyze succeeding reactions.

Every species manufactures proteins unique to that species. The information for synthesizing the unique proteins is located in the nucleus of the cell. The so-called **genetic code** specifies the amino acid sequence in proteins. Hence, the genetic code regulates the chemistry taking place within a cell. Proteins also can serve as a reserve source of energy for the cell. When the amino group is removed from an amino acid, the resulting compound is energy rich.

Nucleic acids

Like proteins, nucleic acids are very large molecules. The nucleic acids are composed of smaller units called **nucleotides.** Each nucleotide contains a carbohydrate molecule, a phosphate group, and a nitrogen-containing molecule that because of its properties is a **nitrogenous base.**

Living organisms have two important nucleic acids. One type is deoxyribonucleic acid, or DNA. The other is **ribonucleic acid,** or **RNA.** DNA is found primarily in the nucleus of the cell, while RNA is found in both the nucleus and the *cytoplasm,* a semi-liquid substance that composes the foundation of the cell (see Chapter 3).

DNA and RNA differ from one another in their components. DNA contains the carbohydrate deoxyribose, while RNA has ribose. In addition, DNA contains the base thymine, while RNA has uracil. The structure of DNA and its importance in cell life is explored in Chapter 10.

Chapter Checkout

Q&A

1. Which of the following is not a component of an atom?

    a. neutron
    b. proton
    c. electron
    d. element

2. Which of the following is not an example of a carbohydrate?

    a. maltose
    b. glycogen
    c. lipid
    d. glucose

3. The building blocks of proteins are _____ ?
   a. DNA
   b. amino acids
   c. enzymes
   d. peptides

4. True or False: The structure of DNA contains the sugar *ribose*.

Answers: 1. d **2.** c **3.** b **4.** False

Chapter 3

THE BIOLOGY OF CELLS

Chapter Check-In

❑ Differentiating prokaryotic and eukaryotic cells

❑ Recognizing the organelles of living cells

❑ Moving materials through the plasma membrane

One of the basic tenets of biology is that all living things are composed of cells. Some organisms consist of a single cell, while others have multiple cells organized into tissues and tissues organized into organs. In many living things, organs function together as an organ system. However, even in these complex organisms, the basic biology revolves around the activities of the cell.

One of the first scientists to observe cells was the Englishman Robert Hooke. In the mid 1600s, Hooke examined a thin slice of cork through the newly invented microscope. The microscopic compartments in the cork impressed him and reminded him of rooms in a monastery, known as cells. He therefore referred to the units as **cells.** Later in that century, Anton Van Leeuwenhoek, a Dutch merchant, made further observations of plant, animal, and microorganism cells. In 1838, the German botanist Matthias Schleiden proposed that all plants are composed of cells. A year later, his colleague, the anatomist Theodore Schwann, concluded that all animals are also composed of cells. In 1858, the biologist Rudolf Virchow proposed that all living things are made of cells and that all cells arise from preexisting cells. These premises have come down to us as the *cell theory.*

The Structure of Prokaryote and Eukaryote Cells

During the 1950s, scientists developed the concept that all organisms may be classified as **prokaryotes** or **eukaryotes.** The cells of all prokaryotes and eukaryotes possess two basic features: a plasma membrane and cytoplasm. However, the cells of prokaryotes are simpler than those of eukaryotes. For example, prokaryotic cells lack a nucleus, while eukaryotic cells have a nucleus. Prokaryotic cells lack internal cellular bodies (organelles), while eukaryotic cells possess them. Examples of prokaryotes are bacteria and cyanobacteria (formerly known as **blue-green algae**). Examples of eukaryotes are protozoa, fungi, plants, and animals.

Plasma membrane

All prokaryote and eukaryote cells have plasma membranes. The **plasma membrane** (also known as the *cell membrane*) is the outermost cell surface, which separates the cell from the external environment. The plasma membrane is composed primarily of proteins and lipids, especially phospholipids. The lipids occur in two layers (a *bilayer*). Proteins embedded in the bilayer appear to float within the lipid, so the membrane is constantly in flux. The membrane is therefore referred to as a *fluid mosaic structure.* Within the fluid mosaic structure, proteins carry out most of the membrane's functions.

The "Movement through the Plasma Membrane" section later in this chapter describes the process by which materials pass between the interior and exterior of a cell.

Cytoplasm and organelles

All prokaryote and eukaryote cells also have **cytoplasm** (or **cytosol**), a semiliquid substance that composes the foundation of a cell. Essentially, cytoplasm is the gel-like material enclosed by the plasma membrane.

Within the cytoplasm of eukaryote cells are a number of membrane-bound bodies called **organelles** ("little organs") that provide a specialized function within the cell.

One example of an organelle is the **endoplasmic reticulum (ER).** The endoplasmic reticulum is a series of membranes extending throughout the cytoplasm of eukaryotic cells. In some places, the ER is studded with submicroscopic bodies called **ribosomes.** This type of ER is referred to as **rough ER.** In other places, there are no ribosomes. This type of ER is called **smooth ER.** The ER is the site of protein synthesis in a cell. Within the ribosomes, amino acids are actually bound together to form proteins. *Cisternae* are spaces within the folds of the ER membranes.

Another organelle is the **Golgi body** (also called the **Golgi apparatus**). The Golgi body is a series of flattened sacs, usually curled at the edges. In the Golgi body, the cell's proteins and lipids are processed and packaged before being sent to their final destination. To accomplish this function, the outermost sac of the Golgi body often bulges and breaks away to form droplike vesicles known as *secretory vesicles.*

An organelle called the *lysosome* (see Figure 3-1) is derived from the Golgi body. It is a droplike sac of **enzymes** in the cytoplasm. These enzymes are used for digestion within the cell. They break down particles of food taken into the cell and make the products available for use. Enzymes are also contained in a cytoplasmic body called the **peroxisome.**

Figure 3-1   The components of an idealized eukaryotic cell. A cell such as this probably does not exist, but the diagram shows the relative sizes and locations of the cell parts.

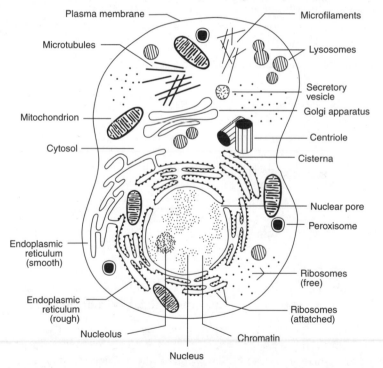

The organelle that releases quantities of energy to form ATP (adenosine triphosphate) is the **mitochondrion** (the plural form is *mitochondria*). Because mitochondria are involved in energy release and storage, they are called the "powerhouses of the cells."

Green plant cells contain organelles known as **chloroplasts,** which function in the process of photosynthesis. Within chloroplasts, energy from the sun is absorbed and transformed into the energy of carbohydrate molecules. Plant cells specialized for photosynthesis contain large numbers of chloroplasts, which are green because the chlorophyll pigments within the chloroplasts are green. Leaves of a plant contain numerous chloroplasts. Plant cells not specializing in photosynthesis (for example, root cells) have few chloroplasts and are not green.

An organelle found in mature plant cells is a large, fluid-filled central vacuole. The vacuole may occupy more than 75 percent of the plant cell. In the vacuole, the plant stores nutrients as well as toxic wastes. Pressure within the growing vacuole may cause the cell to swell.

An organelle called the **cytoskeleton** is an interconnected system of fibers, threads, and interwoven molecules that give structure to the cell. The main components of the cytoskeleton are microtubules, microfilaments, and intermediate filaments. All are assembled from subunits of protein.

The **centriole** organelle is a cylinder-like structure that occurs in pairs. Centrioles function in cell division.

Many cells have specialized cytoskeletal structures called flagella and cilia. *Flagella* are long, hairlike organelles that extend from the cell, permitting it to move. In prokaryotic cells, such as bacteria, the flagella rotate like the propeller of a motorboat. In eukaryotic cells, such as certain protozoa and sperm cells, the flagella whip about and propel the cell. *Cilia* are shorter and more numerous than flagella. In moving cells, the cilia wave in unison and move the cell forward. Paramecium is a well-known ciliated protozoan. Cilia are also found on the surface of several types of cells, such as those that line the human respiratory tract.

Nucleus

Prokaryotic cells lack a **nucleus;** the word *prokaryotic* means "primitive nucleus." Eukaryotic cells, on the other hand, have a distinct nucleus.

The nucleus of eukaryotic cells is composed primarily of protein and **deoxyribonucleic acid,** or **DNA.** The DNA is organized into linear units called **chromosomes,** also known as **chromatin** when the linear units are not obvious. Functional segments of the chromosomes are referred to as **genes.** Approximately 100,000 genes are located in the nucleus of all human cells. Nuclear proteins belong to a class of proteins called **histones.** The chromosome is coiled around the histones.

The *nuclear envelope,* an outer membrane, surrounds the nucleus of a eukaryotic cell. The nuclear envelope is a double membrane, consisting of two lipid layers (similar to the plasma membrane). Pores in the nuclear envelope allow the internal nuclear environment to communicate with the cell cytoplasm.

Within the nucleus are two or more dense organelles referred to as **nucleoli** (the singular form is *nucleolus*). In nucleoli, submicroscopic particles known as *ribosomes* are assembled before their passage out of the nucleus into the cytoplasm.

Although prokaryotic cells have no nucleus, they do have DNA. The DNA exists freely in the cytoplasm as a closed loop. It has no protein to support it and no membrane covering it. A bacterium typically has a single looped chromosome with about 4,000 genes.

Cell wall

Many kinds of prokaryotes and eukaryotes contain a structure outside the cell membrane called the *cell wall.* With only a few exceptions, all bacteria have thick, rigid cell walls that give them their shape. Among the eukaryotes, fungi and plants have cell walls. Cell walls are not identical in these organisms, however. In fungi, the cell wall contains a polysaccharide called *chitin*. Plant cells, in contrast, have no chitin; their cell walls are composed exclusively of the polysaccharide cellulose.

Cell walls provide support and help cells resist mechanical pressures, but they are not solid, so materials are able to pass through rather easily. Cell walls are not selective devices, as plasma membranes are.

Movement through the Plasma Membrane

In order for the cell cytoplasm to communicate with the external environment, materials must be able to move through the plasma membrane. This movement occurs through several mechanisms.

Diffusion

One method of movement through the membrane is **diffusion.** Diffusion is the movement of molecules from a region of higher concentration to one of lower concentration. This movement occurs because the molecules are constantly colliding with one another. The net movement of the molecules is away from the region of high concentration to the region of low concentration.

Diffusion is a random movement of molecules down the pathway called the **concentration gradient.** Molecules are said to move down the concentration gradient because they move from a region of higher concentration to a region of lower concentration. A drop of dye placed in a beaker of water illustrates diffusion as the dye molecules spread out and color the water.

Osmosis

Another method of movement across the membrane is osmosis. **Osmosis** is the movement of water from a region of higher concentration to one of lower concentration. Osmosis often occurs across a membrane that is semipermeable. A semipermeable membrane lets only certain molecules pass through while keeping other molecules out. Osmosis is really a type of diffusion involving only water molecules.

Facilitated diffusion

A third mechanism for movement across the plasma membrane is **facilitated diffusion.** Certain proteins in the membrane assist facilitated diffusion by permitting only certain molecules to pass across the membrane. The proteins encourage movement in the direction that diffusion would normally take place, from a region with a higher concentration of molecules to a region of lower concentration.

Active transport

A fourth method for movement across the membrane is **active transport.** When active transport is taking place, a protein moves a certain material across the membrane from a region of lower concentration to a region of higher concentration. Because this movement is happening against the concentration gradient, the cell must expend energy that is usually derived from a substance called adenosine triphosphate or ATP (see Chapter 4). An example of active transport occurs in human nerve cells. Here, sodium ions are constantly transported out of the cell into the external fluid bathing the cell, a region of high concentration of sodium. (This transport of sodium sets up the nerve cell for the impulse that will occur within it later.)

Endocytosis

The final mechanism for movement across the plasma membrane is **endocytosis,** a process in which a small patch of plasma membrane encloses particles or tiny volumes of fluid that are at or near the cell surface. The membrane enclosure then sinks into the cytoplasm and pinches off from the membrane, forming a vesicle that moves into the cytoplasm. When the vesicle contains particulate matter, the process is called **phagocytosis.** When the vesicle contains droplets of fluid, the process is called **pinocytosis.** Along with the other mechanisms for transport across the plasma membrane, endocytosis ensures that the internal cellular environment will be able to exchange materials with the external environment and that the cell will continue to thrive and function.

Chapter Checkout

Q&A

1. The nucleus contains which of the following?

   a. cytoplasm
   b. DNA
   c. endoplasmic reticulum
   d. Golgi bodies

2. Energy, or ATP, is produced in what organelle?

   a. nucleus
   b. ribosome
   c. mitochondria
   d. vacuole

3. The cytoskeletal organelle is composed of each of the following structures, except _____.

   a. intermediate filaments
   b. microtubules
   c. cilia
   d. microfilaments

4. Movement across the plasma membrane from lower to higher concentration at the expense of energy is an example of which of the following processes?

   a. diffusion
   b. endocytosis
   c. facilitated diffusion
   d. active transport

Answers: 1. b **2.** c **3.** c **4.** d

Chapter 4

CELLS AND ENERGY

Chapter Check-In

❑ Observing the laws of thermodynamics

❑ Catalyzing chemical reactions

❑ Defining the energy unit of a cell

Life can exist only where molecules and cells remain organized. All cells need energy to maintain organization. Physicists define energy as the ability to do work; in this case, the work is the continuation of life itself.

The behavior of energy has been expressed in terms of reliable observations known as the *laws of thermodynamics*. There are two such laws. The first law of thermodynamics states that energy can neither be created nor destroyed. This law implies that the total amount of energy in a closed system (for example, the universe) remains constant. Energy neither enters nor leaves a closed system.

Within a closed system, energy can change, however. For instance, the chemical energy in gasoline is released when the fuel combines with oxygen and a spark ignites the mixture within a car's engine. The gasoline's chemical energy is changed into heat energy, sound energy, and the energy of motion.

The second law of thermodynamics states that the amount of *available* energy in a closed system is decreasing constantly. Energy becomes unavailable for use by living things because of **entropy,** which is the degree of disorder or randomness of a system. The entropy of any closed system is constantly increasing. In essence, any closed system tends toward disorganization.

Unfortunately, the transfers of energy in living systems are never completely efficient. Every body movement, every thought, and every chemical reaction in the cells involves a shift of energy and a measurable loss of energy in the process. For this reason, considerably more energy must be taken into the system than is necessary to carry out the actions of life.

Chemical Reactions

Most chemical compounds do not combine with one another automatically, nor do chemical compounds break apart automatically. The great majority of the chemical reactions that occur within living things must be energized. This means that the atoms of a molecule must be separated by energy put into the system. The energy forces apart the atoms in the molecules and allows the reaction to take place.

To initiate a chemical reaction, a type of "spark," referred to as the *energy of activation,* is needed. For example, hydrogen and oxygen can combine to form water at room temperature, but the reaction requires activation energy.

Any chemical reaction in which energy is released is called an **exergonic reaction.** In an exergonic chemical reaction, the products end up with less energy than the reactants. Other chemical reactions are **endergonic reactions.** In endergonic reactions, energy is obtained and trapped from the environment. The products of endergonic reactions have more energy than the reactants taking part in the chemical reaction. For example, plants carry out the process of photosynthesis in which they trap energy from the sun to form carbohydrates (see Chapter 5).

The activation energy needed to spark an exergonic or endergonic reaction can be heat energy or chemical energy. Reactions that require activation energy can also proceed in the presence of biological catalysts. *Catalysts* are substances that speed up chemical reactions but remain unchanged themselves. Catalysts work by lowering the required amount of activation energy for the chemical reaction. For example, hydrogen and oxygen combine with one another in the presence of platinum. In this case, platinum is the catalyst. In biological systems, the most common catalysts are protein molecules called **enzymes.** Enzymes are absolutely essential if chemical reactions are to occur in cells.

Enzymes

The chemical reactions in all cells of living things operate in the presence of biological catalysts called **enzymes.** Because a particular enzyme catalyzes only one reaction, there are thousands of different enzymes in a cell catalyzing thousands of different chemical reactions. The substance changed or acted on by an enzyme is its **substrate.** The products of a chemical reaction catalyzed by an enzyme are *end products.*

All enzymes are composed of proteins. (Proteins are chains of amino acids; see Chapter 2.) When an enzyme functions, a key portion of the enzyme called the **active site** interacts with the substrate. The active site closely matches the molecular configuration of the substrate. After this interaction has taken place, a change in shape in the active site places a physical stress on the substrate. This physical stress aids the alteration of the substrate and produces the end products. During the time the active site is associated with the substrate, the combination is referred to as the *enzyme-substrate complex.* After the enzyme has performed its work, the product or products drift away. The enzyme is then free to function in another chemical reaction.

Enzyme-catalyzed reactions occur extremely fast. They happen about a million times faster than uncatalyzed reactions. With some exceptions, the names of enzymes end in "-ase." For example, the enzyme that breaks down hydrogen peroxide to water and hydrogen is catalase. Other enzymes include amylase, hydrolase, peptidase, and kinase.

The rate of an enzyme-catalyzed reaction depends on a number of factors, such as the concentration of the substrate, the acidity and temperature of the environment, and the presence of other chemicals. At higher temperatures, enzyme reactions occur more rapidly, but only up to a point. Because enzymes are proteins, excessive amounts of heat can change their structures, rendering them inactive. An enzyme altered by heat is said to be *denatured.*

Enzymes work together in metabolic pathways. A *metabolic pathway* is a sequence of chemical reactions occurring in a cell. A single enzyme-catalyzed reaction may be one of multiple reactions in a metabolic pathway. Metabolic pathways may be of two general types: catabolic and anabolic. Catabolic pathways involve the breakdown or digestion of large, complex molecules. The general term for this process is **catabolism.** Anabolic pathways involve the synthesis of large molecules, generally by joining smaller molecules together. The general term for this process is **anabolism.**

Many enzymes are assisted by chemical substances called **cofactors.** Cofactors may be ions or molecules associated with an enzyme and required in order for a chemical reaction to take place. Ions that might operate as cofactors include those of iron, manganese, and zinc. Organic molecules acting as cofactors are referred to as **coenzymes.** Examples of coenzymes are NAD and FAD (see the "ATP Production" section later in the chapter).

Adenosine Triphosphate (ATP)

The chemical substance that serves as the currency of energy in a cell is **adenosine triphosphate (ATP).** ATP is referred to as currency because it can be "spent" in order to make chemical reactions occur. The more energy required for a chemical reaction, the more ATP molecules must be spent.

Virtually all forms of life use ATP, a nearly universal molecule of energy transfer. The energy released during catabolic reactions is stored in ATP molecules. In addition, the energy trapped in anabolic reactions (such as photosynthesis) is trapped in ATP molecules.

An ATP molecule consists of three parts. One part is a double ring of carbon and nitrogen atoms called *adenine.* Attached to the adenine molecule is a small five-carbon carbohydrate called *ribose.* Attached to the ribose molecule are three phosphate units linked together by covalent bonds.

The covalent bonds that unite the phosphate units in ATP are high-energy bonds. When an ATP molecule is broken down by an enzyme, the third (terminal) phosphate unit is released as a phosphate group, which is an ion. When this happens, approximately 7.3 kilocalories of energy are released. (A kilocalorie equals 1,000 calories.) This energy is made available to do the work of the cell.

The adenosine triphosphatase enzyme accomplishes the breakdown of an ATP molecule. The products of ATP breakdown are **adenosine diphosphate (ADP)** and a **phosphate ion.** Adenosine diphosphate and the phosphate ion can be reconstituted to form ATP, much like a battery can be recharged. To accomplish this, synthesis energy must be available. This energy can be made available in the cell through two extremely important processes: photosynthesis (see Chapter 5) and cellular respiration (see Chapter 6).

ATP Production

ATP is generated from ADP and phosphate ions by a complex set of processes occurring in the cell. These processes depend on the activities of a special group of cofactors called coenzymes. Three important coenzymes are: **nicotinamide adenine dinucleotide (NAD); nicotinamide adenine dinucleotide phosphate (NADP); and flavin adenine dinucleotide (FAD).**

Both NAD and NADP are structurally similar to ATP. Both molecules have a nitrogen-containing ring called nicotinic acid, which is the chemically active part of the coenzymes. In FAD, the chemically active portion is the flavin group. The vitamin riboflavin is used in the body to produce this flavin group.

All coenzymes perform essentially the same work. During the chemical reactions of metabolism, coenzymes accept electrons and pass them on to other coenzymes or other molecules. The removal of electrons or protons from a coenzyme is *oxidation.* The addition of electrons to a molecule is *reduction.* Therefore, the chemical reactions performed by coenzymes are called *oxidation-reduction reactions.*

The oxidation-reduction reactions performed by the coenzymes and other molecules are essential to the energy metabolism of the cell. Other molecules participating in this energy reaction are called **cytochromes.** Together with the coenzymes, cytochromes accept and release electrons in a system referred to as the *electron transport system.* The passage of energy-rich electrons among cytochromes and coenzymes drains the energy from the electrons to form ATP from ADP and phosphate ions.

The actual formation of ATP molecules requires a complex process referred to as **chemiosmosis.** Chemiosmosis involves the creation of a steep proton (hydrogen ion) gradient. This gradient occurs between the membrane-bound compartments of the mitochondria of all cells and the chloroplasts of plant cells. A gradient is formed when large numbers of protons (hydrogen ions) are pumped into the membrane-bound compartments of the mitochondria. The protons build up dramatically within the compartment, finally reaching an enormous number. The energy released from the electrons during the electron transport system pumps the protons.

After large numbers of protons have gathered within the compartments of mitochondria and chloroplasts, they suddenly reverse their directions and escape back across the membranes and out of the compartments. The escaping protons release their energy in this motion. This energy is used by enzymes to unite ADP with phosphate ions to form ATP. The energy is trapped in the high-energy bond of ATP by this process, and the ATP molecules are made available to perform cell work. The movement of protons is chemiosmosis because it is a movement of chemicals (in this case protons) across a semipermeable membrane. Because chemiosmosis occurs in mitochondria and chloroplasts, these organelles play an essential role in the cell's energy metabolism. Chapter 5 explains how energy is trapped in the chloroplasts in plants, while Chapter 6 explains how energy is released in the mitochondria of plant and animal cells.

Chapter Checkout

Q&A

1. The initiation of a chemical reaction requires _____.

   a. exergonic

   b. energy of activation

   c. a catalyst

2. The rate of enzymatic reactions depends on each of the following conditions, except _____.

   a. nutrients

   b. temperature

   c. pH

3. True or False: Adenosine triphosphate (ATP) molecules are used to drive chemical reactions.

4. Each of the following molecules are types of cofactors used to generate ATP, except _____.

   a. NAD

   b. FAD

   c. NADII

Answers: 1. b. **2.** a. **3.** True. **4.** c.

Chapter 5

PHOTOSYNTHESIS

Chapter Check-In

❑ Capturing sunlight for energy

❑ Absorbing light into a photosystem

❑ Activating electrons to create energy

A great variety of living things on earth, including all green plants, synthesize their foods from simple molecules, such as carbon dioxide and water. For this process, the organisms require energy, and that energy is derived from sunlight.

Figure 5-1 shows the energy relationships in living cells. Light energy is captured in the chloroplast of plant cells and used to synthesize glucose molecules, shown as $C_6H_{12}O_6$. In the process, oxygen (O_2) is released as a waste product. The glucose and oxygen are then used in the mitochondrion of the plant and animal cell, and the energy is released and used to fuel the synthesis of ATP from ADP and P. In the reaction, CO_2 and water are released in the mitochondrion to be reused in photosynthesis in the chloroplast.

Figure 5-1 Energy relationships in living cells.

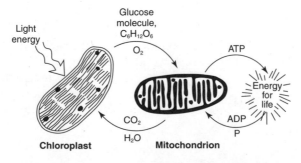

The process of utilizing energy to synthesize carbohydrate molecules is referred to as *photosynthesis*. Photosynthesis is actually two separate processes. In the first process, energy-rich electrons flow through a series of coenzymes and other molecules. This electron energy is trapped. During the trapping process, adenosine triphosphate (ATP) molecules and molecules of nicotinamide adenine dinucleotide phosphate hydrogen (NADPH) are formed. Both ATP and NADPH are rich in energy. These molecules are used in the second half of the process, where carbon dioxide molecules are bound into carbohydrates to form organic substances such as glucose.

Chloroplast

The organelle in which photosynthesis occurs (in the leaves and green stems of plants) is called the **chloroplast.** Chloroplasts are relatively large organelles, containing a watery, protein-rich fluid called *stroma*. The stroma contains many small structures composed of membranes that resemble stacks of coins. Each stack is a *granum* (the plural form is *grana*). Each membrane in the stack is a **thylakoid.** Within the thylakoid membranes (or *thylakoids*) of the granum, many of the reactions of photosynthesis take place. The thylakoids are somewhat similar to the cristae of mitochondria (see Chapter 6).

Photosystems

Pigment molecules organized into photosystems capture sunlight in the chloroplast. **Photosystems** are clusters of light-absorbing pigments with some associated molecules—proton (hydrogen ion) pumps, enzymes, coenzymes, and cytochromes (see Chapter 4). Each photosystem contains about 200 molecules of a green pigment called **chlorophyll** and about 50 molecules of another family of pigments called *carotenoids*. In the reaction center of the photosystem, the energy of sunlight is converted to chemical energy. The center is sometimes called a *light-harvesting antenna*.

There are two photosystems within the thylakoid membranes, designated *photosystem I* and *photosystem II*. The reaction centers of these photosystems are P700 and P680, respectively. The energy captured in these reaction centers drives chemiosmosis, and the energy of chemiosmosis stimulates ATP production in the chloroplasts.

The Process of Photosynthesis

The process of photosynthesis is conveniently divided into two parts: the energy-fixing reaction (also called the light reaction) and the carbon-fixing reaction (also called the light-independent reaction, or the dark reaction).

Energy-fixing reaction

The energy-fixing reaction of photosynthesis begins when light is absorbed in photosystem II in the thylakoid membranes. The energy of the sunlight, captured in the P680 reaction center, activates electrons to jump out of the chlorophyll molecules in the reaction center. These electrons pass through a series of cytochromes in the nearby electron-transport system.

After passing through the electron transport system, the energy-rich electrons eventually enter photosystem 1. Some of the energy of the electron is lost as the electron moves along the chain of acceptors, but a portion of the energy pumps protons across the thylakoid membrane, and this pumping sets up the potential for chemiosmosis.

The spent electrons from P680 enter the P700 reaction center in photosystem I. Sunlight now activates the electrons, which receive a second boost out of the chlorophyll molecules. There they reach a high energy level. Now the electrons progress through a second electron transport system, but this time there is no proton pumping. Rather, the energy reduces NADP. This reduction occurs as two electrons join NADP and energize the molecule. Because NADP acquires two negatively charged electrons, it attracts two positively charged protons to balance the charges. Consequently, the NADP molecule is reduced to NADPH, a molecule that contains much energy.

Because electrons have flowed out of the P680 reaction center, the chlorophyll molecules are left without a certain number of electrons. Electrons secured from water molecules replace these electrons. Each split water molecule releases two electrons that enter the chlorophyll molecules to replace those lost. The split water molecules also release two protons that enter the cytoplasm near the thylakoid and are available to increase the chemiosmotic gradient.

The third product of the disrupted water molecules is oxygen. Two oxygen atoms combine with one another to form molecular oxygen, which is given off as the byproduct of photosynthesis; it fills the atmosphere and is used by all oxygen-breathing organisms, including plant and animal cells.

What has been described above are the *noncyclic energy-fixing reactions* (see Figure 5-2). Certain plants are also known to participate in *cyclic energy-fixing reactions*. These reactions involve only photosystem I and the P700 reaction center. Excited electrons leave the reaction center, pass through coenzymes of the electron transport system, and then follow a special pathway back to P700. Each electron powers the proton pump and encourages the transport of a proton across the thylakoid membrane. This process enriches the proton gradient and eventually leads to the generation of ATP.

Figure 5-2   The energy-fixing reactions of photosynthesis.

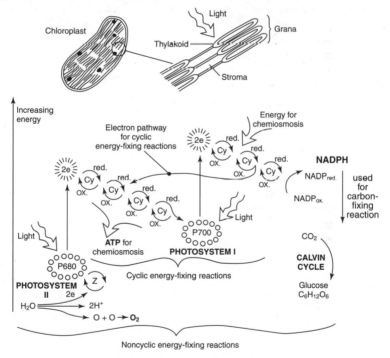

ATP production in the energy-fixing reactions of photosynthesis occurs by the process of **chemiosmosis** (explained in Chapter 4). Essentially, this process consists of a rush of protons across a membrane (the thylakoid membrane, in this case), accompanied by the synthesis of ATP molecules. Biochemists have calculated that the proton concentration on one side of the thylakoid is 10,000 times that on the opposite side of the membrane.

In photosynthesis, the protons pass back across the membranes through channels lying alongside sites where enzymes are located. As the protons pass through the channels, the energy of the protons is released to form high-energy ATP bonds. ATP is formed in the energy-fixing reactions along with the NADPH formed in the main reactions. Both ATP and NADPH provide the energy necessary for the synthesis of carbohydrates that occurs in the second major set of events in photosynthesis.

Carbon-fixing reaction

Glucose and other carbohydrates are synthesized in the carbon-fixing reaction of photosynthesis, often called the *Calvin cycle* for Melvin Calvin, who performed much of the biochemical research (see Figure 5-3). This phase of photosynthesis occurs in the stroma of the plant cell.

Figure 5-3   A carbon-fixing reaction, also called the Calvin cycle.

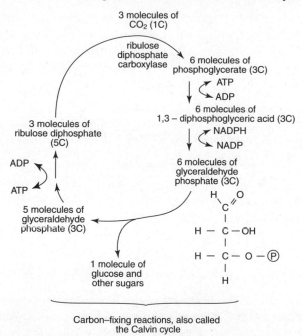

In the carbon-fixing reaction, an essential material is carbon dioxide, which is obtained from the atmosphere. The carbon dioxide is attached to a five-carbon compound called *ribulose diphosphate*. Ribulose diphosphate carboxylase catalyzes this reaction.

After carbon dioxide has been joined to ribulose diphosphate, a six-carbon product forms, which immediately breaks into two three-carbon molecules called *phosphoglycerate*. Each phosphoglycerate molecule converts to another organic compound, but only in the presence of ATP. The ATP used is the ATP synthesized in the energy-fixing reaction. The organic compound formed converts to still another organic compound using the energy present in NADPH. Again, the energy-fixing reaction provides the essential energy. The organic compounds that result each consist of three carbon atoms. Eventually, the compounds interact with one another and join to form a single molecule of six-carbon glucose. This process also generates additional molecules of ribulose diphosphate to participate in further carbon-fixing reactions.

Glucose can be stored in plants in several ways. In some plants, the glucose molecules are joined to one another to form starch molecules. Potato plants, for example, store starch in tubers (underground stems). In some plants, glucose converts to fructose (fruit sugar), and the energy is stored in this form. In still other plants, fructose combines with glucose to form sucrose, commonly known as table sugar. The energy is stored in carbohydrates in this form. Plant cells obtain energy for their activities from these molecules. Animals use the same forms of glucose by consuming plants and delivering the molecules to their cells.

All living things on earth depend in some way on photosynthesis. It is the main mechanism for bringing the energy of sunlight into living systems and making that energy available for the chemical reactions taking place in cells.

Chapter Checkout

Q&A

1. The coenzyme used to generate ATP during photosynthesis is called _____.

   a. NAD
   b. NADP
   c. NADPH

2. Photosynthesis takes place in which of the following organelles?

   a. chlorophyll
   b. chloroplast
   c. stroma

3. The energy fixing reactions generate _____ , whereas the carbon-fixing reactions produce _____ .

   a. ATP, CO_2
   b. $C_6H_{12}O_6$, CO_2
   c. ATP, $C_6H_{12}O_6$

4. True or False: Production of $C_6H_{12}O_6$ during photosynthesis occurs independent of light.

Answers: 1. c **2.** b **3.** c **4.** True

Chapter 6

CELLULAR RESPIRATION

Chapter Check-In

❑ Breaking down glucose during glycolysis

❑ Moving through the Krebs cycle

❑ Transporting electrons and pumping protons

❑ Metabolizing glucose by fermentation

Organisms, such as plants, can trap the energy in sunlight through photosynthesis (see Chapter 5) and store it in the chemical bonds of carbohydrate molecules. The principal carbohydrate formed through photosynthesis is **glucose.** Other types of organisms, such as animals, fungi, protozoa, and a large portion of the bacteria, are unable to perform this process. Therefore, these organisms must rely on the carbohydrates formed in plants to obtain the energy necessary for their metabolic processes.

Animals and other organisms obtain the energy available in carbohydrates through the process of **cellular respiration.** Cells take the carbohydrates into their cytoplasm, and through a complex series of metabolic processes, they break down the carbohydrates and release the energy. The energy is generally not needed immediately; rather it is used to combine adenosine diphosphate (ADP) with phosphate ions to form adenosine triphosphate (ATP) molecules. The ATP can then be used for processes in the cells that require energy, much as a battery powers a mechanical device.

During the process of cellular respiration, carbon dioxide is given off. This carbon dioxide can be used by plant cells during photosynthesis to form new carbohydrates. Also in the process of cellular respiration, oxygen gas is required to serve as an acceptor of electrons. This oxygen is identical to the oxygen gas given off during photosynthesis. Thus, there is an interrelationship between the processes of photosynthesis and cellular respiration, namely the entrapment of energy available in sunlight and the provision of the energy for cellular processes in the form of ATP.

The overall mechanism of cellular respiration involves four processes: glycolysis, in which glucose molecules are broken down to form pyruvic acid molecules; the Krebs cycle, in which pyruvic acid is further broken down and the energy in its molecule is used to form high-energy compounds, such as nicotinamide adenine dinucleotide (NADH); the electron transport system, in which electrons are transported along a series of coenzymes and cytochromes and the energy in the electrons is released; and chemiosmosis, in which the energy given off by electrons pumps protons across a membrane and provides the energy for ATP synthesis. The general chemical equation for cellular respiration is:

$$C_6H_{12}O_6 + 6\ O_2 \rightarrow 6\ H_2O + 6CO_2 + energy$$

Figure 6-1 provides an overview of cellular respiration. Glucose is converted to acetyl-CoA in the cytoplasm, and then the Krebs cycle proceeds in the mitochondrion. Electron transport and chemiosmosis result in energy release and ATP synthesis.

Glycolysis

Glycolysis is the process in which one glucose molecule is broken down to form two molecules of pyruvic acid. The glycolysis process is a multistep metabolic pathway that occurs in the cytoplasm of animal cells, plant cells, and the cells of microorganisms. At least six enzymes operate in the metabolic pathway.

In the first and third steps of the pathway, ATP energizes the molecules. Thus, two ATP molecules must be expended in the process. Further along in the process, the six-carbon glucose molecule converts into intermediary compounds and then is split into two three-carbon compounds. The latter undergo additional conversions and eventually form pyruvic acid at the conclusion of the process.

During the latter stages of glycolysis, four ATP molecules are synthesized using the energy given off during the chemical reactions. Thus, four ATP molecules are synthesized and two ATP molecules are used during glycolysis, for a net gain of two ATP molecules.

Figure 6-1   An overview of cellular respiration.

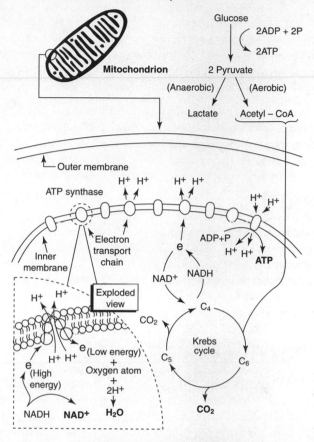

Another reaction during glycolysis yields enough energy to convert NAD to NADH (plus a hydrogen ion). The reduced coenzyme (NADH) will later be used in the electron transport system, and its energy will be released. During glycolysis, two NADH molecules are produced.

Because glycolysis does not use any oxygen, the process is considered to be anaerobic. For certain anaerobic organisms, such as some bacteria and fermentation yeasts, glycolysis is the sole source of energy.

Glycolysis is a somewhat inefficient process because much of the cellular energy remains in the two molecules of pyruvic acid that are created. Interestingly, this process is somewhat similar to a reversal of photosynthesis (see Chapter 5).

Krebs Cycle

Following glycolysis, the mechanism of cellular respiration involves another multistep process—the Krebs cycle, which is also called the citric acid cycle or the tricarboxylic acid cycle. The **Krebs cycle** uses the two molecules of pyruvic acid formed in glycolysis and yields high-energy molecules of NADH and flavin adenine dinucleotide (FADH), as well as some ATP.

The Krebs cycle occurs in the mitochondrion of a cell (see Figure 6-1). This sausage-shaped organelle possesses inner and outer membranes and, therefore, an inner and outer compartment. The inner membrane is folded over itself many times; the folds are called *cristae*. They are somewhat similar to the thylakoid membranes in chloroplasts (see Chapter 5). Located along the cristae are the important enzymes necessary for the proton pump and for ATP production.

Prior to entering the Krebs cycle, the pyruvic acid molecules are altered. Each three-carbon pyruvic acid molecule undergoes conversion to a substance called acetyl-coenzyme A, or acetyl-CoA. During the process, the pyruvic acid molecule is broken down by an enzyme, one carbon atom is released in the form of carbon dioxide, and the remaining two carbon atoms are combined with a coenzyme called coenzyme A. This combination forms acetyl-CoA. In the process, electrons and a hydrogen ion are transferred to NAD to form high-energy NADH.

Acetyl-CoA now enters the Krebs cycle by combining with a four-carbon acid called oxaloacetic acid. The combination forms the six-carbon acid called citric acid. Citric acid undergoes a series of enzyme-catalyzed conversions. The conversions, which involve up to ten chemical reactions, are all brought about by enzymes. In many of the steps, high-energy electrons are released to NAD. The NAD molecule also acquires a hydrogen ion and becomes NADH. In one of the steps, FAD serves as the electron acceptor, and it acquires two hydrogen ions to become $FADH_2$. Also, in one of the reactions, enough energy is released to synthesize a molecule of ATP. Because for each glucose molecule there are two pyruvic acid molecules entering the system, two ATP molecules are formed.

Also during the Krebs cycle, the two carbon atoms of acetyl-CoA are released, and each forms a carbon dioxide molecule. Thus, for each acetyl-CoA entering the cycle, two carbon dioxide molecules are formed. Two acetyl-CoA molecules enter the cycle, and each has two carbon atoms, so four carbon dioxide molecules will form. Add these four molecules to the two carbon dioxide molecules formed in the conversion of pyruvic acid to acetyl-CoA, and it adds up to six carbon dioxide molecules. These six CO_2 molecules are given off as waste gas in the Krebs cycle. They represent the six carbons of glucose that originally entered the process of glycolysis.

At the end of the Krebs cycle, the final product is oxaloacetic acid. This is identical to the oxaloacetic acid that begins the cycle. Now the molecule is ready to accept another acetyl-CoA molecule to begin another turn of the cycle. All told, the Krebs cycle forms (per two molecules of pyruvic acid) two ATP molecules, ten NADH molecules, and two $FADH_2$ molecules. The NADH and the $FADH_2$ will be used in the electron transport system.

Electron Transport System

The electron transport system occurs in the cristae of the mitochondria, where a series of cytochromes (cell pigments) and coenzymes exist. These cytochromes and coenzymes act as carrier molecules and transfer molecules. They accept high-energy electrons and pass the electrons to the next molecule in the system. At key proton-pumping sites, the energy of the electrons transports protons across the membrane into the outer compartment of the mitochondrion.

Each NADH molecule is highly energetic, which accounts for the transfer of six protons into the outer compartment of the mitochondrion. Each $FADH_2$ molecule accounts for the transfer of four protons. The flow of electrons is similar to that taking place in photosynthesis. Electrons pass from NAD to FAD, to other cytochromes and coenzymes, and eventually they lose much of their energy. In cellular respiration, the final electron acceptor is an oxygen atom. In their energy-depleted condition, the electrons unite with an oxygen atom. The electron–oxygen combination then reacts with two hydrogen ions (protons) to form a water molecule (H_2O)

The role of oxygen in cellular respiration is substantial. As a final electron receptor, it is responsible for removing electrons from the system. If oxygen were not available, electrons could not be passed among the coenzymes, the energy in electrons could not be released, the proton pump could not be established, and ATP could not be produced. In humans, breathing is the essential process that brings oxygen into the body for delivery to the cells to participate in cellular respiration.

Chemiosmosis

The actual production of ATP in cellular respiration takes place through the process of chemiosmosis (see Chapter 4). Chemiosmosis involves the pumping of protons through special channels in the membranes of mitochondria from the inner to the outer compartment. The pumping establishes a proton gradient. After the gradient is established, protons pass down the gradient through particles designated F1. In these particles, the energy of the protons generates ATP, using ADP and phosphate ions as the starting points.

The energy production of cellular respiration is substantial. Most biochemists agree that 36 molecules of ATP can be produced for each glucose molecule during cellular respiration as a result of the Krebs cycle reactions, the electron transport system, and chemiosmosis. Also, two ATP molecules are produced through glycolysis, so the grand total is 38 molecules of ATP. These ATP molecules may then be used in the cell for its needs. However, the ATP molecules cannot be stored for long periods of time, so cellular respiration must constantly continue in order to regenerate the ATP molecules as they are used. Each ATP molecule is capable of releasing 7.3 kilocalories of energy per mole.

Fermentation

Fermentation is an anaerobic process in which energy can be released from glucose even though oxygen is not available. **Fermentation** occurs in yeast cells, and a form of fermentation takes place in bacteria and in the muscle cells of animals.

In yeast cells (the yeast used for baking and producing alcoholic beverages), glucose can be metabolized through cellular respiration as in other cells. When oxygen is lacking, however, glucose is still metabolized to pyruvic acid via glycolysis. The pyruvic acid is converted first to acetaldehyde and then to ethyl alcohol. The net gain of ATP to the yeast cell is two molecules—the two molecules of ATP normally produced in glycolysis.

Yeasts are able to participate in fermentation because they have the necessary enzyme to convert pyruvic acid to ethyl alcohol. This process is essential because it removes electrons and hydrogen ions from NADH during glycolysis. The effect is to free the NAD so it can participate in future reactions of glycolysis. The net gain to the yeast cell of two ATP molecules permits it to remain alive for some time. However, when the percentage of ethyl alcohol reaches approximately 15 percent, the alcohol kills the yeast cells.

Yeast is used both in bread and alcohol production. Alcohol fermentation is the process that yields beer, wine, and other spirits. The carbon dioxide given off during fermentation supplements the carbon dioxide given off during the Krebs cycle and causes bread to rise.

In muscle cells, another form of fermentation takes place. When muscle cells contract too frequently (as in strenuous exercise), they rapidly use up their oxygen supply. As a result, the electron transport system and Krebs cycle slow considerably, and ATP production is slowed. However, muscle cells have the ability to produce a small amount of ATP through glycolysis in the absence of oxygen. The muscle cells convert glucose to pyruvic acid. Then an enzyme in the muscle cells converts the pyruvic acid to lactic acid. As in the yeast, this reaction frees up the NAD while providing the cells with two ATP molecules from glycolysis. Eventually, however, the lactic acid buildup causes intense fatigue, and the muscle cell stops contracting.

Chapter Checkout

Q&A

1. True or False: Cellular respiration takes place in chloroplasts.

2. Prior to entering the Krebs cycle, two molecules of _____ are broken down to form _____.
   a. $C_6H_{12}O_6$, pyruvate
   b. $C_6H_{12}O_6$, acetyl-CoA
   c. pyruvate, acetyl-CoA
   d. Acetyl-CoA, $C_6H_{12}O_6$

3. For one molecule of glucose, _____ molecules of ATP are formed during the Krebs cycle.
   a. 2
   b. 1
   c. 4
   d. 8

4. For each glucose molecule, _____ molecules of ATP are formed during respiration.

    a. 24

    b. 38

    c. 36

    d. 12

Answers: 1. False. **2.** c **3.** a **4.** b

Chapter 7

MITOSIS AND CELL REPRODUCTION

Chapter Check-In

❏ Reproducing within the nucleus

❏ Cycling through cell division

❏ Separating nuclear components

Adistinguishing feature of a living thing is that it reproduces independent of other living things. This reproduction occurs at the cellular level. In certain parts of the body, such as along the gastrointestinal tract, the cells reproduce often. In other parts of the body, such as in the nervous system, the cells reproduce less frequently. With the exception of only a few kinds of cells, such as red blood cells (which lack nuclei), all cells of the human body reproduce.

Cell Nucleus

In eukaryotic cells (see Chapter 3), the structure and contents of the nucleus are of fundamental importance to an understanding of cell reproduction. The nucleus contains the hereditary material of the cell assembled into chromosomes. In addition, the nucleus usually contains one or more prominent **nucleoli** (dense bodies that are the site of ribosome synthesis).

The nucleus is surrounded by a nuclear envelope consisting of a double membrane that is continuous with the endoplasmic reticulum. Transport of molecules between the nucleus and cytoplasm is accomplished through a series of nuclear pores lined with proteins that facilitate the passage of molecules out of and into the nucleus. The proteins provide a certain measure of selectivity in the passage of molecules across the nuclear membrane.

The nuclear material consists of deoxyribonucleic acid (DNA) organized into long strands. The strands of DNA are composed of nucleotides bonded to one another by covalent bonds. DNA molecules are extremely long relative to the cell; indeed, the length of a chromosome may be hundreds of times the diameter of its cell. However, in the chromosome, the **DNA** is condensed and packaged with protein into manageable bodies. The mass of DNA material and its associated protein is **chromatin.**

To form chromatin, the DNA molecule is wound around globules of a protein called **histone.** The units formed in this way are *nucleosomes.* Millions of nucleosomes are connected by short stretches of histone protein much like beads on a string. The configuration of the nucleosomes in a coil causes additional coiling of the DNA and the eventual formation of the chromosome.

Cell Cycle

The **cell cycle** involves many repetitions of cellular growth and reproduction. With few exceptions (for example, red blood cells), all the cells of living things undergo a cell cycle.

The cell cycle is generally divided into two phases: interphase and mitosis. During **interphase,** the cell spends most of its time performing the functions that make it unique. *Mitosis* is the phase of the cell cycle during which the cell divides into two daughter cells.

Interphase

The interphase stage of the cell cycle includes three distinctive parts: the G_1 phase, the S phase, and the G_2 phase. The **G1 phase** follows mitosis and is the period in which the cell is synthesizing its structural proteins and enzymes to perform its functions. For example, a pancreas cell in the G_1 phase will produce and secrete insulin, a muscle cell will undergo the contractions that permit movement, and a salivary gland cell will secrete salivary enzymes to assist digestion. During the G_1 phase, each chromosome consists of a single molecule of DNA and its associated histone protein. In human cells, there are 46 chromosomes per cell (except in sex cells with 23 chromosomes and red blood cells with no nucleus and hence no chromosomes).

During the **S phase** of the cell cycle, the DNA within the nucleus replicates. During this process, each chromosome is faithfully copied, so by the end of the S phase, two DNA molecules exist for each one formerly present in the G_1 phase. Human cells contain 92 chromosomes per cell in the S phase.

In the G_2 phase, the cell prepares for mitosis. Proteins organize themselves to form a series of fibers called the *spindle,* which is involved in chromosome movement during mitosis. The spindle is constructed from amino acids for each mitosis, and then taken apart at the conclusion of the process. Spindle fibers are composed of microtubules.

Mitosis

The term *mitosis* is derived from the Latin stem *mito,* meaning "threads." When mitosis was first described a century ago, scientists had seen "threads" within cells, so they gave the name mitosis to the process of "thread movement." During mitosis, the nuclear material becomes visible as threadlike chromosomes. The chromosomes organize in the center of the cell, and then they separate, and 46 chromosomes move into each new cell that forms.

Mitosis is a continuous process, but for convenience in denoting which portion of the process is taking place, scientists divide mitosis into a series of phases: prophase, metaphase, anaphase, telophase, and cytokinesis (see Figure 7-1):

Figure 7-1   The process of mitosis, in which the chromosomes of a cell duplicate and pass into two daughter cells.

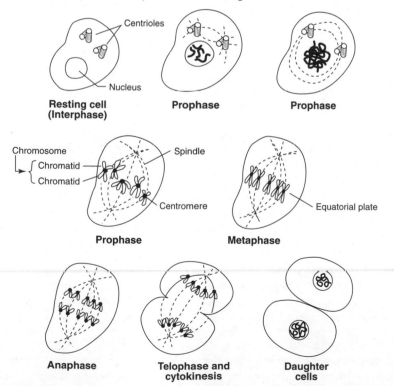

■ **Prophase:** Mitosis begins with the condensation of the chromosomes to form visible threads in the phase called prophase. Two copies of each chromosome exist; each one is a **chromatid.** Two chromatids are joined to one another at a region called the **centromere.** As prophase unfolds, the chromatids become visible in pairs, the spindle fibers form, the nucleoli disappear, and the nuclear envelope dissolves.

In animal cells during prophase, microscopic bodies called the **centrioles** begin to migrate to opposite sides of the cell. When the centrioles reach the poles of the cell, they produce, and are then surrounded by, a series of radiating microtubules called an *aster.* Centrioles and asters are not present in most plant or fungal cells.

As prophase continues, the chromatids attach to spindle fibers that extend out from opposite poles of the cell. The spindle fibers attach at the region of the centromere at a structure called the **kinetochore,** a region of DNA that has remained undivided. Eventually, all pairs of chromatids reach the center of the cell, a region called the *equatorial plate.*

■ **Metaphase:** Metaphase is the stage of mitosis in which the pairs of chromatids line up on the equatorial plate. This region is also called the metaphase plate. In a human cell, 92 chromosomes in 46 pairs align at the equatorial plate. Each pair is connected at centromere, where the spindle fiber is attached (more specifically at the kinetochore). At this point, the DNA at the kinetochore duplicates, and the two chromatids become completely separate from one another.

■ **Anaphase:** At the beginning of anaphase, the chromatids move apart from one another. The chromatids are **chromosomes** after the separation. Each chromosome is attached to a spindle fiber, and the members of each chromosome pair are drawn to opposite poles of the cell by the spindle fibers. During anaphase, the chromosomes can be seen moving. They take on a rough V shape because of their midregion attachment to the spindle fibers. The movement toward the poles is accomplished by several mechanisms, such as an elongation of the spindle fibers, which results in pushing the poles apart.

The result of anaphase is an equal separation and distribution of the chromosomes. In humans cells, a total of 46 chromosomes move to each pole as the process of mitosis continues.

■ **Telophase:** In telophase, the chromosomes finally arrive at the opposite poles of the cell. The distinct chromosomes begin to fade from sight as masses of chromatin are formed again. The events of telophase are essentially the reverse of those in prophase. The spindle is dismantled and its amino acids are recycled, the nucleoli reappear, and the nuclear envelope is reformed.

■ **Cytokinesis:** Cytokinesis is the process in which the cytoplasm divides and two separate cells form. In animal cells, cytokinesis begins with the formation of a furrow in the center of the cell. With the formation of the furrow, the cell membrane begins to pinch into the cytoplasm, and the formation of two cells begins. This process is often referred to as *cell cleavage.* Microfilaments contract during cleavage and assist the division of the cell into two daughter cells.

In plant cells, cytokinesis occurs by a different process because a rigid cell wall is involved. Cleavage does not take place in plant cells. Rather, a new cell wall is assembled at the center of the cell, beginning with vesicles formed from the Golgi body (see Chapter 3). As the vesicles join, they form a double membrane called the *cell plate.* The cell plate forms in the middle of the cytoplasm and grows outward to fuse with the cell membrane. The cell plate separates the two daughter cells. As cell wall material is laid down, the two cells move apart from one another to yield two new daughter cells.

Mitosis serves several functions in living cells. In many simple organisms, it is the method for asexual reproduction (for example, in the cells of a fungus). In multicellular organisms, mitosis allows the entire organism to grow by forming new cells and replacing older cells. In certain species, mitosis is used to heal wounds or regenerate body parts. It is the universal process for cell division.

Chapter Checkout

Q&A

1. Red blood cells contain _____ chromosomes.

  a. 46
  b. 23
  c. 0

2. Mitosis is divided into each of the following phases, except _____ .

   a. interphase

   b. prophase

   c. cytokinesis

3. _____ attach to a region of the centromere called _____ .

   a. spindle fibers, kinetochore

   b. centriole, kinetochore

   c. aster, centriole

Answers: 1. c **2.** a **3.** a

Chapter 8

MEIOSIS AND GAMETE FORMATION

Chapter Check-In

❑ Generating haploid cells

❑ Dividing by meiosis

❑ Crossing over

Most plant and animal cells are diploid. The term **diploid** is derived from the Greek *diplos,* meaning "double" or "two"; the term implies that the cells of plants and animals have two sets of chromosomes. In human cells, for example, 46 chromosomes are organized in 23 pairs. Hence, human cells are diploid in that they have two sets of 23 chromosomes per set.

During sexual reproduction, the sex cells of parent organisms unite with one another and form a fertilized egg cell. In this situation, each sex cell is a **gamete.** The gametes of human cells are **haploid,** from the Greek *haplos,* meaning "single." This term implies that each gamete contains a single set of chromosomes—23 chromosomes in humans. When the human gametes unite with one another, the original diploid condition of 46 chromosomes is reestablished. Mitosis then brings about the development of the diploid cell into an organism.

The process by which the chromosome number is halved during gamete formation is **meiosis.** In meiosis, a cell containing the diploid number of chromosomes is converted into four cells, each having the haploid number of chromosomes. In human cells, for instance, a reproductive cell containing 46 chromosomes yields four cells, each with 23 chromosomes.

Meiosis occurs by a series of steps that resemble the steps of mitosis. Two major phases of meiosis occur: meiosis I and meiosis II. During meiosis I, a single cell divides into two. During meiosis II, those two cells each divide again. The same demarcating phases of mitosis take place in meiosis I and meiosis II.

As shown in Figure 8-1, first, the chromosomes of a cell duplicate and pass into two cells. The chromosomes of the two cells then separate and pass into four daughter cells. The parent cell has two sets of chromosomes and is diploid, while the daughter cells have a single set of chromosomes each and are haploid. Synapsis and crossing over occur in the Prophase I stage.

Figure 8-1 The process of meiosis, in which four haploid cells are formed.

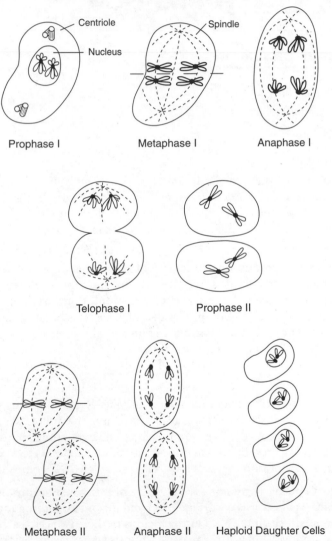

The members of each chromosome pair within a cell are called *homologous chromosomes.* Homologous chromosomes are similar but not identical. They may carry different versions of the same genetic information. For instance, one homologous chromosome may carry the information for blond hair while the other homologous chromosome may carry the information for black hair.

Meiosis

As a cell prepares to enter meiosis, each of its chromosomes has duplicated, as in mitosis. Each chromosome thus consists of two chromatids.

Meiosis I

At the beginning of meiosis 1, a human cell contains 46 chromosomes, or 92 chromatids (the same number as during mitosis). Meiosis I proceeds through the following phases:

- **Prophase I:** Prophase I is similar in some ways to prophase in mitosis. The chromatids shorten and thicken and become visible under a microscope. An important difference, however, is that a process called synapsis occurs. A second process called crossing over also takes place during prophase 1.

  During prophase 1, the two homologous chromosomes come near each other. Because each homologous chromosome consists of two chromatids, there are actually four chromatids aligned next to one another. This combination of four chromatids is called a *tetrad,* and the coming together is the process called *synapsis.*

  After synapsis has taken place, the process of crossing over occurs. In this process, segments of DNA from one chromatid in the tetrad pass to another chromatid in the tetrad. These exchanges of chromosomal segments occur in a complex and poorly understood manner. They result in a genetically new chromatid. Crossing over is an important driving force of evolution. After crossing over has taken place, the four chromatids of the tetrad are genetically different from the original four chromatids.

- **Metaphase I:** In metaphase I of meiosis, the tetrads align on the equatorial plate (as in mitosis). The centromeres attach to spindle fibers, which extend from the poles of the cell. One centromere attaches per spindle fiber.

- **Anaphase I:** In anaphase 1, the homologous chromosomes separate. One homologous chromosome (consisting of two chromatids) moves to one side of the cell, while the other homologous chromosome (consisting of two chromatids) moves to the other side of the cell. The result is that 23 chromosomes (each consisting of two chromatids) move to one pole, and 23 chromosomes (each consisting of two chromatids) move to the other pole. Essentially, the chromosome number of the cell is halved. For this reason the process is a reduction-division.

- **Telophase I:** In telophase I of meiosis, the nucleus reorganizes, the chromosomes become chromatin, and a cytoplasmic division into two cells takes place. This process occurs differently in plant and animal cells, just as in mitosis. Each daughter cell (with 23 chromosomes each consisting of two chromatids) then enters interphase, during which there is no duplication of the DNA. The interphase period may be brief or very long, depending on the species of organism.

Meiosis II

Meiosis II is the second major subdivision of meiosis. It occurs in essentially the same way as mitosis. In meiosis II, a cell containing 46 chromatids undergoes division into two cells, each with 23 chromosomes. Meiosis II proceeds through the following phases:

- **Prophase II:** Prophase II is similar to the prophase of mitosis. The chromatin material condenses, and each chromosome contains two chromatids attached by the centromere. The 23 chromatid pairs, a total of 46 chromatids, then move to the equatorial plate.

- **Metaphase II:** In metaphase II of meiosis, the 23 chromatid pairs gather at the center of the cell prior to separation. This process is identical to metaphase in mitosis.

- **Anaphase II:** During anaphase II of meiosis, the centromeres divide, and the 46 chromatids become known as 46 chromosomes. Then the 46 chromosomes separate from one another. Spindle fibers move one chromosome from each pair to one pole of the cell and the other member of the pair to the other pole. In all, 23 chromosomes move to each pole. The forces and attachments that operate in mitosis also operate in anaphase 11.

■ **Telophase II:** During telophase II, the chromosomes gather at the poles of the cells and become indistinct. Again, they form a mass of chromatin. The nuclear envelope develops, the nucleoli reappear, and the cells undergo cytokinesis as in mitosis.

During meiosis II, each cell containing 46 chromatids yields two cells, each with 23 chromosomes. Originally, there were two cells that underwent meiosis II; therefore, the result of meiosis II is four cells, each with 23 chromosomes. Each of the four cells is haploid; that is, each cell contains a single set of chromosomes.

The 23 chromosomes in the four cells from meiosis are not identical because crossing over has taken place in prophase 1. The crossing over yields variation so that each of the four resulting cells from meiosis differs from the other three. Thus, meiosis provides a mechanism for producing variations in the chromosomes. Also, it accounts for the formation of four haploid cells from a single diploid cell.

Meiosis in Humans

In humans, meiosis is the process by which sperm cells and egg cells are produced. In the male, meiosis takes place after puberty. Diploid cells within the testes undergo meiosis to produce haploid **sperm** cells with 23 chromosomes. A single diploid cell yields four haploid sperm cells through meiosis.

In females, meiosis begins during the fetal stage when a series of diploid cells enter meiosis 1. At the conclusion of meiosis 1, the process comes to a halt, and the cells gather in the ovaries. At puberty, meiosis resumes. One cell at the end of meiosis I enters meiosis II each month. The result of meiosis II is a single egg cell per cycle (the other meiotic cells disintegrate). Each egg cell contains 23 chromosomes and is haploid.

The union of the egg cell and the sperm cell leads to the formation of a fertilized egg cell with 46 chromosomes, or 23 pairs. Fertilization restores the diploid number of chromosomes. The fertilized egg cell, a diploid, is a **zygote.** Further divisions of the zygote by mitosis eventually yield a complete human being.

Chapter Checkout

Q&A

1. As a cell enters meiosis I, the cell contains _____ chromosomes, or _____ chromatids.

a. 46, 92
b. 23, 46
c. 46, 23
d. 92, 46

2. Each diploid human cell contains _____ pairs of homologous chromosomes.

a. 46
b. 23
c. 96
d. 36

3. A reproductive cell, containing 46 chromosomes will yield _____ cells with _____ chromosomes each after meiosis.

a. 2, 23
b. 4, 46
c. 4, 23
d. 8, 23

4. True or False: As a cell exits meiosis I and prepares to enter meiosis II, the cell reproduces its DNA.

Answers: 1. a **2.** b **3.** c **4.** False

Chapter 9

CLASSICAL (MENDELIAN) GENETICS

Chapter Check-In

❑ Introducing Gregor Mendel

❑ Predicting inheritance patterns

❑ Defining the principles of genetics

Genetics is the study of how genes bring about characteristics, or traits, in living things and how those characteristics are inherited. Genes are portions of DNA molecules that determine characteristics of living things. Through the processes of meiosis and reproduction, genes are transmitted from one generation to the next.

The Augustinian monk Gregor Mendel developed the science of genetics. Mendel performed his experiments in the 1860s and 1870s, but the scientific community did not accept his work until early in the twentieth century. Because the principles established by Mendel form the basis for genetics, the science is often referred to as *Mendelian genetics*. It is also called *classical genetics* to distinguish it from another branch of biology known as molecular genetics (see Chapter 10).

Mendel believed that factors pass from parents to their offspring, but he did not know of the existence of DNA. Modern scientists accept that **genes** are composed of segments of DNA molecules that control discrete hereditary characteristics.

Most complex organisms have cells that are diploid. **Diploid** cells have a double set of chromosomes, one from each parent. For example, human cells have a double set of chromosomes consisting of 23 pairs, or a total of 46 chromosomes. In a diploid cell, there are two genes for each characteristic. In preparation for sexual reproduction, the diploid number of chromosomes is reduced to a haploid number. That is, diploid cells are reduced to cells that have a single set of chromosomes. These haploid cells are **gametes,** or sex cells, and they are formed through meiosis (see Chapter 8). When gametes come together in sexual reproduction, the diploid condition is reestablished.

The offspring of sexual reproduction obtain one gene of each type from each parent. The different forms of a gene are called **alleles.** In humans, for instance, there are two alleles for earlobe construction. One allele is for earlobes that are attached, while the other allele is for earlobes that hang free. The type of earlobe a person has is determined by the alleles inherited from the parents.

The set of all genes that specify an organism's traits is known as the organism's **genome.** The genome for a human cell consists of about 100,000 genes. The gene composition of a living organism is its **genotype.** For a person's earlobe shape, the genotype may consist of two genes for attached earlobes, or two genes for free earlobes, or one gene for attached and one gene for free earlobes.

The expression of the genes is referred to as the **phenotype** of a living thing. If a person has attached earlobes, the phenotype is "attached earlobes." If the person has free earlobes, the phenotype is "free earlobes." Even though three genotypes for earlobe shape are possible, only two phenotypes (attached earlobes and free earlobes) are possible.

The two paired alleles in an organism's genotype may be identical, or they may be different. An organism's condition is said to be **homozygous** when two identical alleles are present for a particular characteristic. In contrast, the condition is said to be **heterozygous** when two different alleles are present for a particular characteristic. In a homozygous individual, the alleles express themselves. In a heterozygous individual, the alleles may interact with one another, and in many cases, only one allele is expressed.

When one allele expresses itself and the other does not, the one expressing itself is the **dominant** allele. The overshadowed allele is the **recessive** allele. In humans, the allele for free earlobes is the dominant allele. If this allele is present with the allele for attached earlobes, the allele for free earlobes expresses itself, and the phenotype of the individual is "free earlobes." Dominant alleles always express themselves, while recessive alleles express themselves only when two recessive alleles exist together in an individual. Thus, a person having free earlobes can have one dominant allele or two dominant alleles, while a person having attached earlobes must have two recessive alleles.

Inheritance Patterns

Mendel was the first scientist to develop a method for predicting the outcome of inheritance patterns. He performed his work with pea plants, studying seven traits: plant height, pod shape, pod color, seed shape, seed color, flower color, and flower location. Pea plants pollinate themselves. Therefore, over many generations, pea plants develop individuals that are homozygous for particular characteristics. These populations are known as *pure lines*.

In his work, Mendel took pure-line pea plants and cross-pollinated them with other pure-line pea plants. He called these plants the *parent generation*. When Mendel crossed pure-line tall plants with pure-line short plants, he discovered that all the plants resulting from this cross were tall. He called this generation the *F1 generation* (first filial generation). Next, Mendel crossed the offspring of the F_1 generation tall plants among themselves to produce a new generation called the *F2 generation* (second filial generation). Among the plants in this generation, Mendel observed that three-fourths of the plants were tall and one-fourth of the plants were short.

Mendel's laws of genetics

Mendel conducted similar experiments with the other pea plant traits. Over many years, he formulated several principles that are known today as Mendel's laws of genetics. His laws include the following:

1. **Mendel's law of dominance:** When an organism has two different alleles for a trait, one allele dominates.

2. **Mendel's law of segregation:** During gamete formation by a diploid organism, the pair of alleles for a particular trait separate, or segregate, during the formation of gametes (as in meiosis).

3. **Mendel's law of independent assortment:** The members of a gene pair separate from one another independent of the members of other gene pairs. (These separations occur in the formation of gametes during meiosis.)

Mendelian crosses

An advantage of genetics is that scientists can predict the probability of inherited traits in offspring by performing a genetic cross (also called a *Mendelian cross*). To predict the possibility of an individual trait, several steps are followed. First, a symbol is designated for each allele in the gene pair. The dominant allele is represented by a capital letter and the recessive allele by the corresponding lowercase letter, such as *E* for free earlobes and *e* for attached earlobes. For a homozygous dominant individual, the genotype would be *EE;* for a heterozygous individual, the genotype would be *Ee;* and for a homozygous recessive person, the genotype would be *ee.*

The next step in performing a genetic cross is determining the genotypes of the parents and the genotype of the gametes. A heterozygous male and a heterozygous female to be crossed have the genotypes of *Ee* and *Ee.* During meiosis, the allele pairs separate. A sperm cell contains either an *E* or an *e,* while the egg cell also contains either an *E* or an *e.*

To continue the genetics problem, a Punnett square is used. A *Punnett square* is a boxed figure used to determine the probability of genotypes and phenotypes in the offspring of a genetic cross. The possible gametes produced by the female are indicated at the top of the square, while the possible gametes produced by the male are indicated at the left side of the square. Figure 9-1 shows the Punnett square for the earlobe example.

Figure 9–1 An example of a Punnett square.

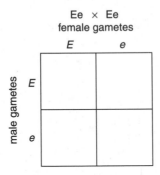

Continuing, all of the possible combinations of alleles are considered. This is done by filling in each square with the alleles above it and at its left. This is done as shown in Figure 9-2.

From the Punnett square, the phenotype of each possible genotype can be determined. For example, the offspring having *EE, Ee,* and *Ee* will have free earlobes. Only the offspring with the genotype *ee* will have attached earlobes. Therefore, the ratio of phenotypes is three with free earlobes to one with attached earlobes (3:1). The ratio of genotypes is 1:2:1 (1 *EE : 2 Ee : 1 ee*).

Figure 9-2 The Punnett square is used to determine the probabilities of the genotypes and phenotypes in the offspring of a genetic cross.

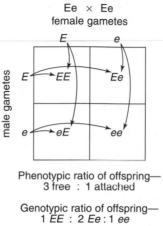

Ee × Ee
female gametes

Phenotypic ratio of offspring—
3 free : 1 attached

Genotypic ratio of offspring—
1 *EE* : 2 *Ee* : 1 *ee*

Principles of Genetics

Mendel's studies have provided scientists with the basis for mathematically predicting the probabilities of genotypes and phenotypes in the offspring of a genetic cross. But not all genetic observations can be explained and predicted based on Mendelian genetics. Other complex and distinct genetic phenomena may also occur. Several complex genetic concepts, described in this section, explain such distinct genetic phenomena as blood types and skin color.

Incomplete dominance

In some allele combinations, dominance does not exist. Instead, the two characteristics blend. In such a situation, both alleles have the opportunity to express themselves. For instance, snapdragon flowers display **incomplete dominance** in their color. There are two alleles for flower color: one for white and one for red. When two alleles for white are present, the plant displays white flowers. When two alleles for red are present, the plant has red flowers. But when one allele for red is present with one allele for white, the color of the snapdragons is pink.

However, if two pink snapdragons are crossed, the phenotype ratio of the offspring is one red, two pink, and one white. These results show that the genes themselves remain independent; only the expressions of the genes blend. If the gene for red and the gene for white actually blended, pure red and pure white snapdragons could not appear in the offspring.

Multiple alleles

In certain cases, more than two alleles exist for a particular characteristic. Even though an individual has only two alleles, additional alleles may be present in the population. This condition is **multiple alleles.**

An example of multiple alleles occurs in blood type. In humans, blood groups are determined by a single gene with three possible alleles: A, B, or O. Red blood cells can contain two antigens, A and B. The presence or absence of these antigens results in four blood types: A, B, AB, and O. If a person's red blood cells have antigen A, the blood type is A. If a person's red blood cells have antigen B, the blood type is B. If the red blood cells have both antigen A and antigen B, the blood type is AB. If the red blood cells have neither antigen A nor antigen B, the blood type is O.

The alleles for type A and type B blood are co-dominant; that is, both alleles are expressed. However, the allele for type O blood is recessive to both type A and type B. Because a person has only two of the three alleles, the blood type varies depending on which two alleles are present. For instance, if a person has the A allele and the B allele, the blood type is AB. If a person has two A alleles, or one A and one O allele, the blood type is A. If a person has two B alleles or one B and one O allele, the blood type is B. If a person has two O alleles, the blood type is O.

Polygenic inheritance

Although many characteristics are determined by alleles at a single place on the chromosome, some characteristics are determined by an interaction of genes on several chromosomes or at several places on one chromosome. This condition is **polygenic inheritance.**

An example of polygenic inheritance is human skin color. Genes for skin color are located in many places, and skin color is determined by which genes are present at these multiple locations. A person with many genes for dark skin will have very dark skin color, and a person with multiple genes for light skin will have very light skin color. Many people have some genes for light skin and some for dark skin, which explains why so many variations of skin color exist. Height is another characteristic probably reflecting polygenic characteristics.

Gene linkage

A chromosome has many thousands of genes; there are an estimated 100,000 genes in the human genome. Inheritance involves the transfer of chromosomes from parent to offspring through meiosis and sexual reproduction. It is common for a large number of genes to be inherited together if they are located on the same chromosome. Genes that are inherited together are said to form a linkage group. The concept of transfer of a linkage group is **gene linkage.**

Gene linkage can show how close two or more genes are to one another on a chromosome. The closer the genes are to each other, the higher the probability that they will be inherited together. Crossing over occurs during meiosis, but genes that are close to each other tend to remain together during crossing over.

Sex linkage

Among the 23 pairs of chromosomes in human cells, one pair is the **sex chromosomes.** (The remaining 22 pairs of chromosomes are referred to as **autosomes.**) The sex chromosomes determine the sex of humans. There are two types of sex chromosomes: the X chromosome and the Y chromosome. Females have two X chromosomes; males have one X and one Y chromosome. Typically, the female chromosome pattern is designated XX, while the male chromosome pattern is XY. Thus, the genotype of the human male would be 44 XY, while the genotype of the human female would be 44 XX (where 44 represents the autosomes).

In humans, the Y chromosome is much shorter than the X chromosome. Because of this shortened size, a number of sex-linked conditions occur. When a gene occurs on an X chromosome, the other gene of the pair probably occurs on the other X chromosome. Therefore, a female usually has two genes for a characteristic. In contrast, when a gene occurs on an X chromosome in a male, there is usually no other gene present on the short Y chromosome. Therefore, in the male, whatever gene is present on the X chromosome will be expressed.

An example of a sex-linked trait is colorblindness. The gene for colorblindness is found on the X chromosome. A woman is rarely colorblind because she usually has a dominant gene for normal vision on one of her X chromosomes. However, a male has the shortened Y chromosome; therefore, he has no gene to offset a gene for colorblindness on the X chromosome. As a result, the gene for colorblindness expresses itself in the male.

Another example of sex-linked inheritance is the blood disease hemophilia. In hemophilia, the blood does not clot normally because an important blood-clotting protein is missing. The gene for hemophilia occurs on the X chromosome. As females have two X chromosomes, one X chromosome usually has the gene for normal blood clotting. Therefore, the female may be a carrier of hemophilia but normally does not express hemophilia. Males have no offsetting gene on the Y chromosome, so the gene for hemophilia expresses itself in the male. This is why most cases of hemophilia occur in males.

Chapter Checkout

Practice Project

1. Create a Punnett Square and determine the genotypes and phenotypes of the resulting offspring if one parent is heterozygous for free earlobes (Ee) and the other parent is homozygous for attached earlobes (ee).

Q&A

2. If both parents are heterozygous for the gene for free earlobes, what will be the ratio of phenotypes of the resulting offspring?

3. Two pink snapdragons were mated to each other. What are the resulting genotypic and phenotypic ratios of the offspring?

Answers: 1. 2 Ee = free earlobes; 2 ee = attached earlobes. **2.** The ratio will be 3:1, 3 free earlobes to 1 attached. **3.** 1:2:1 genotypic ratio (RR = 1, Rr = 2, rr = 1). 1:2:1 phenotypic ratio (red = 1, pink = 2, white = 1).

Chapter 10

GENE EXPRESSION
(MOLECULAR GENETICS)

Chapter Check-In

❑ Discovering the double helix

❑ Replicating the DNA

❑ Synthesizing an RNA transcript

❑ Translating into an amino acid

During the 1950s, a tremendous explosion of biological research occurred, and the methods of gene expression were elucidated. The knowledge generated during this period helped explain how genes function, and it gave rise to the science of molecular genetics. This science is based on the activity of deoxyribonucleic acid (DNA) and how this activity brings about the production of proteins in the cell. Genetic material is packaged into DNA molecules. DNA molecules relay the inherited information to messenger RNA (mRNA) which, in turn, codes for proteins. This chain of command is represented as:

$$DNA \rightarrow mRNA \rightarrow protein$$

The flow of information from DNA to protein is known as the *Central Dogma* of molecular biology.

In 1953, two biochemists, James D. Watson and Francis H.C. Crick, proposed a model for the structure of DNA. (In 1962, they shared a Nobel Prize for their work.) The publication of the structure of DNA opened a new realm of molecular genetics. Its structure provided valuable insight into how genes operate and how DNA can reproduce itself during mitosis, thereby passing on hereditary characteristics. Not only did the new research uncover many of the principles of protein synthesis, but it also gave rise to the science of biotechnology and genetic engineering (see Chapter 11).

DNA Structure

As proposed by Watson and Crick, **deoxyribonucleic acid (DNA)** consists of two long nucleotide chains. The two nucleotide chains twist around one another to form a double helix, a shape resembling a spiral staircase. Weak chemical bonds between the chains hold the two chains of nucleotides to one another.

A **nucleotide** in the DNA chain consists of three parts: a nitrogenous base, a phosphate group, and a molecule of deoxyribose. The **nitrogenous bases** of each nucleotide chain are of two major types: **purines** and **pyrimidines.** Purine bases have two fused rings of carbon and nitrogen atoms, while pyrimidines have only one ring. The two purine bases in DNA are adenine (A) and guanine (G). The pyrimidine bases in DNA are cytosine (C) and thymine (T). Purines and pyrimidine bases are found in both strands of the double helix.

The **phosphate group** of DNA is derived from a molecule of phosphoric acid. The phosphate group connects the deoxyribose molecules to one another in the nucleotide chain. **Deoxyribose** is a five-carbon carbohydrate. The purine and pyrimidine bases are attached to the deoxyribose molecules, and the purine and pyrimidine bases are opposite one another on the two nucleotide chains. Adenine is always opposite thymine and binds to thymine. Guanine is always opposite cytosine and binds to cytosine. Adenine and thymine are said to be complementary, as are guanine and cytosine This is known as the principle of complementary base pairing.

DNA Replication

Before a cell enters the process of mitosis, its DNA replicates itself. Equal copies of the DNA pass into the daughter cells at the end of mitosis. In human cells, this means that 46 chromosomes (or molecules of DNA) replicate to form 92 chromosomes.

The process of DNA replication begins when specialized enzymes pull apart, or "unzip," the DNA double helix (see Figure 10-1). As the two strands separate, the purine and pyrimidine bases on each strand are exposed. The exposed bases then attract their complementary bases.

Deoxyribose molecules and phosphate groups are present in the nucleus. The enzyme **DNA polymerase** joins all the nucleotide components to one another, forming a long strand of nucleotides. Thus, the old strand of DNA directs the synthesis of a new strand of DNA through complementary base pairing. The old strand then unites with the new strand to reform a double helix. This process is called *semiconservative replication* because one of the old strands is conserved in the new DNA double helix.

Figure 10-1   DNA replication. The double helix opens and a complementary strand of DNA is synthesized along each strand.

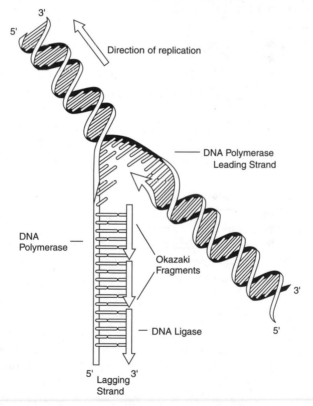

DNA polymerase joins nucleotides in a 5'-3' direction on the leading strand, shown in Figure 10-1. However, DNA polymerase does not elongate a DNA strand in a 3'-5' direction. Therefore, the 3'-5' strand, called the lagging strand, is synthesized in short segments in a 5'-3' direction. These short segments placed on the lagging strand are **Okazaki fragments** and are ultimately joined together by the enzyme DNA ligase to form a new DNA strand.

DNA replication occurs during the S phase of the cell cycle. After replication has taken place, the chromosomal material shortens and thickens. The chromatids appear in the prophase of the next mitosis. The process then continues, and eventually two daughter cells form, each with the identical amount and kind of DNA as the parent cell. The process of DNA replication thus ensures that the molecular material passes to the offspring cells in equal amounts and types.

Protein Synthesis

During the 1950s and 1960s, it became apparent that DNA is essential in the synthesis of proteins. **Proteins** are used in enzymes and as structural materials in cells. Many specialized proteins function in cellular activities. For example, in humans, the hormone insulin and the muscle cell filaments are composed of protein. The hair, skin, and nails of humans are composed of proteins, as are all the hundreds of thousands of enzymes in the body.

The key to a protein molecule is how the amino acids are linked. The sequence of amino acids in a protein is a type of code that specifies the protein and distinguishes one protein from another. A genetic code in the DNA determines this amino acid code. The genetic code consists of the sequence of nitrogenous bases in the DNA. How the nitrogenous base code is translated to an amino acid sequence in a protein is the basis for protein synthesis.

For protein synthesis to occur, several essential materials must be present, such as a supply of the 20 amino acids, which comprise most proteins. Another essential element is a series of enzymes that will function in the process. DNA and another form of nucleic acid called **ribonucleic acid (RNA)** are essential.

RNA is the nucleic acid that carries instructions from the nuclear DNA into the cytoplasm, where protein is synthesized. RNA is similar to DNA, with two exceptions. First, the carbohydrate in RNA is ribose rather than deoxyribose, and second, RNA nucleotides contain the pyrimidine uracil rather than thymine.

Types of RNA

In the synthesis of protein, three types of RNA function. The first type is called *ribosomal RNA (rRNA)*. This form of RNA is used to manufacture ribosomes. **Ribosomes** are ultramicroscopic particles of rRNA and protein. They are the places (the chemical "workbenches") where amino acids are linked to one another to synthesize proteins. Ribosomes may exist along the membranes of the endoplasmic reticulum or in the cytoplasm of the cell (see Chapter 3).

A second important type of RNA is *transfer RNA (tRNA)*. Transfer RNA exists in the cell cytoplasm and carries amino acids to the ribosomes for protein synthesis. When protein synthesis is taking place, enzymes link tRNA molecules to amino acids in a highly specific manner. For example, tRNA molecule X will link only to amino acid X; tRNA Y will link only to amino acid Y.

The third form of RNA is *messenger RNA (mRNA)*. In the nucleus, messenger RNA receives the genetic code in the DNA and carries the code into the cytoplasm where protein synthesis takes place. Messenger RNA is synthesized in the nucleus at the DNA molecules. During the synthesis, the genetic information is transferred from the DNA molecule to the mRNA molecule. In this way, a genetic code can be used to synthesize a protein in a distant location. **RNA polymerase,** an enzyme, accomplishes mRNA, tRNA, and rRNA synthesis.

Transcription

Transcription is one of the first processes in the mechanism of protein synthesis. In transcription, a complementary strand of mRNA is synthesized according to the nitrogenous base code of DNA. To begin, the enzyme RNA polymerase binds to an area of one of the DNA molecules in the double helix. (During transcription, only one DNA strand serves as a template for RNA synthesis. The other DNA strand remains dormant.) The enzyme moves along the DNA strand and "reads" the nucleotides one by one. Similar to the process of DNA replication, the new nucleic acid strand elongates in a 5'-3' direction, as shown in Figure 10-2. The enzyme selects complementary bases from available nucleotides and positions them in an mRNA molecule according to the principle of complementary base pairing. The chain of mRNA lengthens until a "stop" message is received.

Figure 10-2 The process of transcription. The DNA double helix opens and the enzyme RNA polymerase synthesizes a molecule of mRNA according to the base sequence of the DNA template.

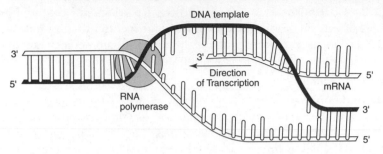

The nucleotides of the DNA strands are read in groups of three. Each group is a *codon*. Thus, a codon may be CGA, or TTA, or GCT, or any other combination of the four bases, depending on their sequence in the DNA strand. Each codon will later serve as a "code word" for an amino acid. First, however, the codons are transcribed to the mRNA molecule. Thus, the mRNA molecule consists of nothing more than a series of codons received from the genetic message in the DNA.

After the "stop" codon is reached, the synthesis of the mRNA comes to an end. The mRNA molecule leaves the DNA molecule, and the DNA molecule rewinds to form a double helix. Meanwhile, the mRNA molecule passes through a pore in the nucleus and proceeds into the cellular cytoplasm where it moves toward the ribosomes.

Translation

The genetic code is transferred to an amino acid sequence in a protein through the **translation** process, which begins with the arrival of the mRNA molecule at the ribosome. While the mRNA was being synthesized, tRNA molecules were uniting with their specific amino acids according to the activity of specific enzymes. The tRNA molecules then began transporting their amino acids to the ribosomes to meet the mRNA molecule.

After it arrives at the ribosomes, the mRNA molecule exposes its bases in sets of three, the codons. Each codon has a complementary codon called an **anticodon** on a tRNA molecule. When the codon of the mRNA molecule complements the anticodon on the tRNA molecule, the latter places the particular amino acid in that position. Then the next codon of the mRNA is exposed, and the complementary anticodon of a tRNA molecule matches with it. The amino acid carried by the second tRNA molecule is positioned next to the first amino acid, and the two are linked. At this point, the tRNA molecules release their amino acids and return to the cytoplasm to link up with new molecules of amino acid.

When it's time for the next amino acid to be positioned in the growing protein, a new codon on the mRNA molecule is exposed, and the complementary three-base anticodon of a tRNA molecule positions itself opposite the codon. This brings another amino acid into position, and that amino acid links to the previous amino acids. The ribosome moves further down the mRNA molecule and exposes another codon, which attracts another tRNA molecule with its anticodon.

One by one, amino acids are added to the growing chain until the ribosome has moved down to the end of the mRNA molecule. Because of the specificity of tRNA molecules for their individual amino acids, and because of the base pairing between codons and anticodons, the sequence of codons on the mRNA molecule determines the sequence of amino acids in the protein being constructed. And because the codon sequence of the mRNA complements the codon sequence of the DNA, the DNA molecule ultimately directs the amino acid sequencing in proteins. The primary "start" codon on an mRNA molecule is AUG, which codes for the amino acid methionine. Therefore, each mRNA transcript begins with the AUG codon, and the resulting peptide begins with methionine.

Figure 10-3 shows that the process of protein synthesis begins with the production of mRNA (upper right). The mRNA molecule proceeds to the ribosome where it meets tRNA molecules carrying amino acids (upper left). The tRNA molecule has a base code that complements the mRNA code and thereby brings a specific amino acid into position. The amino acids join together in peptide bonds (bottom), and the tRNA molecules are released to pick up additional amino acid molecules.

After the protein has been synthesized completely, it is removed from the ribosome for further processing and to perform its function. For example, the protein may be stored in the Golgi body before being released by the cell, or it may be stored in the lysosome as a digestive enzyme. Also, a protein may be used in the cell as a structural component, or it may be released as a hormone, such as insulin. After synthesis, the mRNA molecule breaks up and the nucleotides return to the nucleus. The tRNA molecules return to the cytoplasm to unite with fresh molecules of amino acids, and the ribosome awaits the arrival of a new mRNA molecule.

Figure 10-3   The process of protein synthesis.

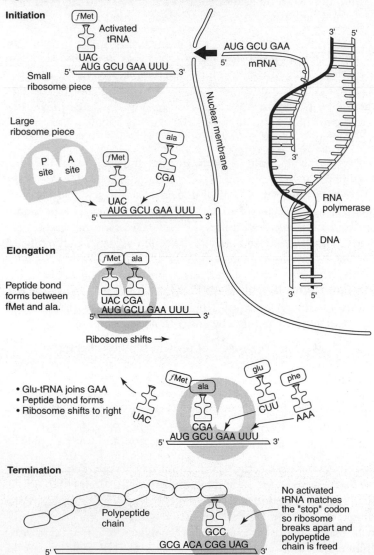

Gene control

The process of protein synthesis does not occur constantly in the cell. Rather, it occurs at intervals followed by periods of genetic "silence." Thus the cell regulates and controls the gene expression process.

The control of gene expression may occur at several levels in the cell. For example, genes rarely operate during mitosis, when the DNA fibers shorten and thicken to form chromatids. The inactive chromatin is compacted and tightly coiled, and this coiling regulates access to the genes.

Other levels of gene control can occur during and after transcription. In transcription, certain segments of DNA can increase and accelerate the activity of nearby genes. After transcription has taken place, the mRNA molecule can be altered to regulate gene activity. For example, researchers have found that an mRNA molecule contains many useless bits of RNA that are removed in the production of the final mRNA molecule. These useless bits of nucleic acid are called *introns.* The remaining pieces of mRNA, called *exons,* are then spliced to form the final mRNA molecule. Thus, through removal of introns and the retention of exons, the cell can alter the message received from the DNA and control gene expression.

The concept of gene control has been researched thoroughly in bacteria. In these microorganisms, genes have been identified as structural genes, regulator genes, and control genes (or control regions). The three units form a functional unit called the *operon.*

The operon has been examined in close detail in certain bacteria. Scientists have found, for example, that certain carbohydrates can induce the presence of the enzymes needed to digest those carbohydrates. When lactose is present, bacteria synthesize the enzyme needed to break down the lactose. Lactose acts as the inducer molecule in the following way: In the absence of lactose, a regulator gene produces a repressor, and the repressor binds to a control region called the operator. This binding prevents the structural genes from encoding the enzyme for lactose digestion. When lactose is present, however, it binds to the repressor and thereby removes the repressor at the operator site. With the operator site free, the structural genes are free to produce their lactose-digesting enzyme.

The operon system in bacteria shows how gene expression can occur in relatively simple cells. The gene is inactive until it is needed and is active when it becomes necessary to produce an enzyme. Other methods of gene control are more complex and are currently being researched.

Chapter Checkout

Q&A

1. What is the mRNA complement, from 5'-3', of the following DNA sequence?

   5'-GATCCGCTAAGGCCT-3'

2. What is the DNA complement to the DNA sequence above?

3. DNA replication takes place during _____ of the cell division cycle.

   a. G_1
   b. metaphase
   c. S phase

4. Which of the following components are not involved in translation?

   a. mRNA
   b. DNA
   c. ribosomes

Answers: 1. 5'-AGGCCUUAGCGGAUC-3'
2. 5'-AGGCCTTAGCGGATC-3' **3.** c **4.** b

Chapter 11

RECOMBINANT DNA AND BIOTECHNOLOGY

Chapter Check-In

❑ Using biotechnology as a tool

❑ Detecting genetic disease

❑ Manufacturing proteins to fight disease

❑ Analyzing a genetic footprint

Biotechnology is an industrial process that uses the scientific research on DNA for practical means. Biotechnology is synonymous with genetic engineering because the genes of an organism are changed during the process. Because the genes are changed, the DNA of the organism is said to be *recombined*. The result of the process is **recombinant DNA.** Recombinant DNA and biotechnology can be used to form proteins not normally produced in a cell, to produce drugs or vaccines, or to promote human health. In addition, bacteria that carry recombinant DNA can be released into the environment under carefully controlled conditions to increase the fertility of the soil, serve as an insecticide, or relieve pollution.

Biotechnology and recombinant DNA can also be used in forensic medicine to "fingerprint" individuals and identify DNA at a crime scene. In addition, transgenic plants and animals are being created. Humans can also have the genes in their cells modified to produce proteins that relieve health-related deficiencies. Finally, at the time of publication, scientists at Celera Genomics and the National Human Genome Research Institute have sequenced a substantial portion of the human genome in an effort to identify genes linked to human disease. The sequenced fragments of DNA are currently not organized into contiguous reading frames. Although a substantial portion of the genome has been sequenced, it may require several more years for all the pieces to come together to form a complete and accurate genetic map of the human genome.

Tools of Biotechnology

The basic process of recombinant DNA technology revolves around DNA activity in the synthesis of protein (see Chapter 10). During this synthesis, DNA provides the genetic code for the placement of amino acids in proteins. By intervening in this process, scientists can change the nature of the DNA, thereby changing the nature of the protein expressed by that DNA. By inserting genes into the genome of an organism, the scientist can induce the organism to produce a protein it does not normally produce.

The technology of recombinant DNA has been made possible in part by extensive research on microorganisms during the last half-century. One important microorganism in recombinant DNA research is *Escherichia coli,* commonly referred to as E. coli. The biochemistry and genetics of E. coli are well known, and its DNA has been isolated and made to accept new genes. The DNA can then be forced into fresh E. coli cells and the bacteria will begin to produce the proteins specified by the foreign genes. Such altered bacteria are said to have been *transformed.*

Knowledge about viruses has also aided the development of DNA technology. **Viruses** are fragments of nucleic acid surrounded by a protein coat. In some cases, viruses attack cells and replicate within the cells, thereby destroying them. In other cases, the viruses enter cells, and their nucleic acid joins with the nucleic acid in the cell nucleus. By attaching DNA to viruses, scientists use viruses to transport foreign DNA into cells and to connect it with the nucleic acid of the cells.

Another common method for inserting DNA into cells is to use **plasmids,** which are small loops of DNA in the cytoplasm of bacterial cells. Working with a plasmid is much easier than working with a chromosome, so plasmids are often the carriers, or **vectors,** of DNA. Plasmids can be isolated, recombined with foreign DNA, then inserted into cells where they multiply as the cells multiply.

Interest in recombinant DNA and biotechnology heightened considerably during the 1960s and 1970s with the discovery of **restriction enzymes.** These enzymes catalyze the opening of a DNA molecule at a "restricted" point, regardless of the source of the DNA. Figure 11-1 shows that a human DNA molecule is opened at a certain site by the restriction enzyme

EcoRl (upper left), and the desired DNA fragment is isolated (lower left). Plasmid DNA is treated with the same enzyme and opened. The DNA fragment is spliced into the plasmid to produce the recombinant DNA molecule.

Figure 11-1 Construction of a recombinant DNA molecule.

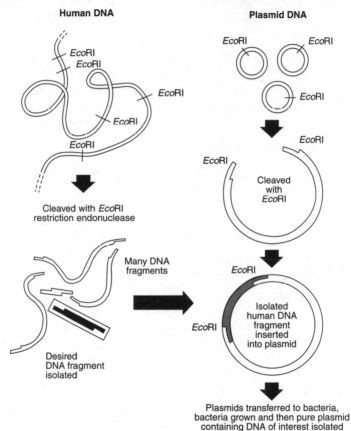

Certain restriction enzymes leave dangling ends of DNA molecules at the point where the DNA is opened. Foreign DNA can therefore be combined with the carrier DNA at this point. An enzyme called DNA ligase forges a permanent link between the dangling ends of the DNA molecules at the point of union.

Recombinant DNA technology is sophisticated and expensive. Genes must be isolated, vectors must be identified, and gene control must be maintained. Stability of the vector within a host cell is important, and the scientist must be certain that nonpathogenic bacteria are used. Cells from mammals can be used to synthesize proteins, but cultivating these cells is difficult. In addition, the proper gene signals must be identified, RNA molecules must be bound to ribosomes, and the presence of introns must be considered. Collecting the gene product and exporting it from the cell are other considerations.

The genes used in DNA technology are commonly obtained from host cells or organisms called *gene libraries*. A gene library is a collection of cells identified as harboring a specific gene. For example, E. coli cells can be stored with the genes for human insulin in their chromosomes.

Pharmaceutical Products

Gene defects in humans can lead to deficiencies in proteins, such as insulin, human growth hormone, and Factor VIII that may result in problems, such as diabetes, dwarfism, and impaired blood clotting, respectively. Proteins for these chemicals can now be replaced by proteins manufactured through biotechnology. For insulin production, two protein chains are encoded by separate genes in plasmids inserted into bacteria. The protein chains are then chemically joined to form the final insulin product. Human growth hormone is also produced within bacteria, but special techniques are used because the bacteria do not usually produce human proteins.

Therapeutic proteins, such as the following, can also be produced by biotechnology:

- **Tissue plasminogen activator (TPA),** a clot-dissolving protein, can now be produced in recombined mammalian cells.

- **Interferon,** an antiviral protein produced in E. coli cells, is currently used to fight certain types of cancers and for certain skin diseases.

- An **antisense molecule** is a molecule of RNA that reacts with and neutralizes the mRNA molecule used in protein synthesis. In doing so, the antisense molecule prevents the synthesis of a protein involved in a specific disease. For example, an antisense molecule can prohibit human host cells from producing key portions of the human immunodeficiency virus (HIV) when infection has occurred.

Vaccines represent another application of recombinant DNA technology. The hepatitis B vaccine now in use is composed of viral proteins manufactured by yeast cells and recombined with viral genes. The vaccine is safe because it contains no viral particles. Experimental vaccines against AIDS are being produced in the same way. One vaccine uses vaccinia (cowpox) virus as a vector. The virus has been combined with genes from several viruses, and it is hoped that injections of the vaccine will stimulate resistance to multiple diseases. Vaccines can also be produced by eliminating certain disease-inducing genes from a pathogen, leading to a harmless organism that will stimulate an immune response.

Diagnostic Testing

Recombinant DNA and biotechnology have opened a new era of diagnostic testing and have made detecting many genetic diseases possible. The basic tool of DNA analyses is a fragment of DNA called the *DNA probe*. A DNA probe is a relatively small, single-stranded fragment of DNA that recognizes and binds to a complementary section of DNA on a larger DNA molecule. The probe mingles within the mixture of DNA molecules and unites with the target DNA much like a left hand unites with a right. After the probe unites with its target, it emits a signal like that from a radioactive isotope to indicate that a reaction has occurred.

To work effectively, a sufficiently large amount of target DNA must be available. To increase the amount of available DNA, a process called the **polymerase chain reaction (PCR)** is used. In a highly automated machine, the target DNA is combined with enzymes, nucleotides, and a primer DNA. In geometric fashion, the enzymes synthesize copies of the target DNA, so that in a few hours billions of molecules of DNA exist where only a few were before.

Using DNA probes and PCR, scientists are now able to detect the DNA associated with HIV (and AIDS). This has yielded a direct test for AIDS that is preferred over the AIDS antibody test. Lyme disease and genetic diseases such as cystic fibrosis, muscular dystrophy, Huntington's disease, and fragile X syndrome can be identified by DNA probes. (Cystic fibrosis is a respiratory disease in which mucus clogs the respiratory passageways and makes breathing difficult; muscular dystrophy is a disorder of the nervous system in which destruction of nerve fibers leads to erratic muscular activity; Huntington's disease is a disease of the nervous system accompanied by erratic movements and nervous degeneration; and fragile X syndrome is a disease of the X chromosome accompanied by a form of mental retardation.)

Segments of DNA called *restriction fragment length polymorphisms* (**RFLPs**) are the objectives of the tests with gene probes. RFLPs are apparently useless bits of DNA located near genes associated with the diseases. By locating RFLPs, the biotechnologist can locate the disease gene. DNA probes also detect microorganisms in the environment and identify viral and bacterial pathogens.

Gene Therapy

Gene therapy is a recombinant DNA process in which cells are taken from the patient, altered by adding genes, and replaced in the patient. The genes then provide the genetic codes for proteins the patient is lacking. Nonreproductive cells are used in gene therapy, so there is no carryover of inserted genes to the next generation.

In the early 1990s, gene therapy was used to correct a deficiency of the enzyme adenosine deaminase (ADA). Blood cells called lymphocytes were removed from the bone marrow of two children, then genes for ADA production were inserted into the cells using viruses as vectors. Finally, the cells were reinfused in the bodies of the children. Once established in the bodies, the gene-altered cells began synthesizing the enzyme ADA. Thus, the deficiency was removed and the disease resolved.

Approximately 2,000 single-gene defects are believed to exist, and patients with these defects may be candidates for gene therapy. A multilayered review system now exists to ensure the safety of gene therapy proposals. Many aspects must be considered before approval is granted for gene therapy experiments.

DNA Fingerprinting

The use of DNA probes and the development of retrieval techniques have made it possible to match DNA molecules to one another for identification purposes. This process has been used in a forensic procedure called **DNA fingerprinting.**

DNA fingerprinting depends on the presence of repeating base sequences that exist in the human genome. The repeating sequences are called restriction fragment length polymorphisms (RFLPs). As the pattern of RFLPs is unique for each individual, it can be used as a molecular fingerprint.

To perform DNA fingerprinting, DNA is obtained from an individual's blood cells, hair fibers, skin fragments, or other tissue. The DNA is then extracted from the cells and digested with enzymes. The resulting fragments are separated by a process called electrophoresis. Electrophoresis is a process in which electrical charges separate DNA fragments according to size. The separated DNA fragments are then detected with DNA probes and used to develop a fingerprint. A statistical evaluation enables the forensic pathologist to compare a suspect's DNA with the DNA recovered at a crime scene and to state with a high degree of certainty (usually 99 percent) that the suspect was at the crime scene.

Searching for DNA

The ability to retrieve DNA from ancient materials and museum specimens has given archaeologists and anthropologists hopes of a glimpse at ancient life. Biochemists have successfully obtained DNA from extinct animals and plants. Evolutionary biologists have used the DNA to draw lineage patterns from the data. This often gives a better understanding of relationships between species. DNA isolated from ancient humans has been used to trace the movements of populations, such as the Anglo-Saxons, as well as to determine whether males were favored over females in certain societies.

Studies have also been performed on human origins by using the DNA found in the mitochondria. All of an offspring's mitochondrial DNA is derived from its mother. Because this DNA represents an unbroken line of genetic information, an analysis of mutation sites in the mitochondrial DNA can conceivably lead one back to the first human female.

DNA and Agriculture

Although plants are more difficult to work with than bacteria, gene insertions can be made into single plant cells. Then the cells can be cultivated to form a mature plant. The major method for inserting genes is through the plasmids of the bacterium called *Agrobacterium tumefaciens.* This bacterium invades plant cells, and its plasmids insert into plant chromosomes carrying the genes for tumor induction. Scientists remove the tumor-inducing genes and obtain a plasmid that unites with the plant cell without causing any harm.

Recombinant DNA and biotechnology have been used to increase the efficiency of plant growth by increasing the efficiency of the plant's ability to fix nitrogen. Scientists have obtained the genes for nitrogen fixation from bacteria and have incorporated those genes into plant cells. The plant cells can then perform a process that normally takes place only in bacteria.

DNA technology has also been used to increase plant resistance to disease by reengineering the plant to produce viral proteins. Also, the genes for an insecticide obtained from a bacterium have been inserted into plants to allow the plants to resist caterpillars and other pests.

One of the first agricultural products of biotechnology was the rot-resistant tomato. This plant was altered by adding a gene that produces an antisense molecule. The antisense molecule inhibits the tomato from producing the enzyme that encourages rotting. Without this enzyme, the tomato can ripen longer on the vine.

Transgenic Animals

Transgenic animals are animals in which one or more genes have been introduced into its nonreproductive cells. The first transgenic animal was produced in 1983 when genes for human growth hormone were introduced into mice.

Transgenic animals can be used to produce valuable products. For example, a transgenic pig has been produced with the ability to synthesize human hemoglobin for use as a blood substitute. Also, a transgenic cow has been bred with the ability to produce human lactoferrin, an iron-building milk protein and a potential antibacterial agent. A transgenic sheep can synthesize a protein that helps emphysema patients breathe easily, and a transgenic goat has been developed to produce a protein needed by cystic fibrosis patients.

Human Genome

In 1990, researchers at Celera Genomics and at the National Human Genome Research Institute began an ambitious endeavor to sequence the entire human genome. In 2000, researchers revealed to the general public that a substantial portion of this work had been completed. This "rough draft" of the human genome is currently in fragments much like pieces of a jigsaw puzzle. Current efforts are underway to match the different pieces to complete the puzzle. Recently, Celera Genomics revealed their startling estimation of the number of human genes to be 30,000. This estimation, based on the sequence data, is substantially below previous predictions. The sequence data has led to the estimation that less than 5% of the human genome actually encodes functional proteins. Once the jigsaw puzzle is completed, the data will undoubtedly help researchers devise new diagnostics and treatments for genetic diseases.

In addition to sequencing the human genome, researchers have sequenced the genomes of Drosophila melanogaster (fruit fly), Arabidopsis thaliana (plant), Saccharomyces cerevisiae (budding yeast), and C. elegans (worm). In addition, mouse, rat, and zebrafish genomes have been sequenced. Not only eukaryotic organisms are useful to the research community. The genome of Plasmodium (the organism that causes malaria) has also been sequenced. The goals of these sequencing projects are to prepare gene linkage maps and physical maps. A gene linkage map is a map that pinpoints the location of genes based on their connection to certain marker gene sequences. A physical map, in comparison, gives the actual number of bases between genes on a chromosome; therefore, it locates the gene of interest more precisely.

Ultimately, scientists hope to learn the actual names and sequences of all 3 billion nitrogenous bases in the human genome. Automation and computerization are essential tools in the sequencing, and the development of the specific technology is underway.

Chapter Checkout

Q&A

1. An enzyme called _____ links the ends of DNA molecules, forming a permanent bond.
   a. restriction
   b. polymerase
   c. ligase
   d. virus

2. A technique used to amplify a specific fragment of DNA is called _____.
   a. RFLP
   b. transgenic
   c. ligation
   d. PCR

3. DNA was extracted from an individual's skin cells. The DNA was digested using restriction enzymes and resolved by electrophoresis. The fragments were analyzed to develop a pattern. This technique is called _____.

a. PCR
b. RFLP
c. gene therapy
d. DNA probe

Answers: 1. c **2.** d **3.** b

Chapter 12

PRINCIPLES OF EVOLUTION

Chapter Check-In

❑ Evolving by natural selection

❑ Validating the theory of evolution

❑ Driving the evolutionary process

Evolution implies a change in one or more characteristics in a population of organisms over a period of time. The concept of evolution is as ancient as Greek writings, where philosophers speculated that all living things are related to one another, although remotely. The Greek philosopher Aristotle perceived a "ladder of life" where simple organisms gradually change to more elaborate forms. Opponents of this concept were led by several theologians who pointed to the biblical account of creation as set forth in the Book of Genesis. One prelate, James Ussher, calculated that creation had taken place on October 26, 4004 B.C., at 9 a.m.

Opponents of the creationist argument were encouraged by geologists who postulated that the earth is far older than 4,004 years. In 1785, James Hutton postulated that the earth was formed by an ancient progression of natural events, including erosion, disruption, and uplift. In the early 1800s, Georges Cuvier suggested that the earth was 6,000 years old, based on his calculations. In 1830, Charles Lyell published evidence pushing the age of the earth back several million years.

Amid the controversy over geology and the age of the earth, French zoologist Jean Baptiste de Lamarck suggested a theory for evolution based on the development of new traits in response to a changing environment. For example, the neck of the giraffe stretched as it reached for food. Lamarck's theory of "use and disuse" gained favor, and his concept of "acquired characteristics" was accepted until the time of Charles Darwin, many years later.

Charles Darwin was the son of an English physician. As a naturalist on the ship *H.M.S. Beagle,* Darwin traveled to remote regions of South America. His observations on this trip led him to develop his own theory of evolution. Darwin was particularly interested in the finches and tortoises of the Galapagos Islands. He pondered how different species of animals could have developed on this remote set of islands 200 miles west of Ecuador.

Darwin returned to England from South America in 1838 and continued to ponder the theory of evolution. He was influenced by Thomas Malthus's *Essay on the Principle of Population.* In his book, Malthus pointed out the human population's continual struggle for survival. Darwin applied this principle to animals and plants, and his theory of evolution began to develop.

In 1858, another English naturalist, Alfred Russell Wallace, developed a concept of evolution similar to Darwin's. Wallace wrote a paper on the subject and corresponded with Darwin. The two men decided to simultaneously present papers on evolution to London's scientific community in 1858. The next year, 1859, Darwin published his famous book, *On the Origin of Species by Means of Natural Selection, or the Preservation of Favoured Races in the Struggle for Life.* The book has become known simply as *The Origin of Species.*

Theory of Evolution

In his book *The Origin of Species,* Darwin presents evidence in a sober manner for his "descent by modification" theory, which has come down to us as the theory of **evolution,** although Darwin avoided the term "evolution." Essentially, Darwin suggested that random variations take place in living things and that some external agent in the environment selects those individuals better able to survive. The method of selecting individuals is known as **natural selection.** The selected individuals pass on their traits to their offspring, and the population continues to evolve.

Two essential points underlie natural selection. First, the genetic variations that take place in living things are random variations. Second, the genetic variations are small and cause little effect relative to a given population. Over time, these small genetic variations lead to the gradual development of a species rather than the sudden development of a species. Darwin proposed that variations appear without direction and without design. He assumed that among inherited traits, some traits were better than others.

If an inherited trait provided an advantage over another, it would provide a reproductive advantage to the bearer of the trait. Thus, if long-necked giraffes could reach food better than short-necked giraffes, the long-necked giraffes would survive, reproduce, and yield a population consisting solely of long-necked giraffes.

As the central concept of Darwin's theory of evolution, natural selection implies that the fittest survive and spread their traits through a population. This concept is referred to as the **survival of the fittest.** The fitness implied is reproductive fitness, that is, the ability to survive in the environment and propagate the species. Natural selection serves as a sieve to remove the unfit from a population and allow the fittest to reproduce and continue the population. Today, scientists know that other factors also influence evolution.

Evidence for Evolution

In his book, Darwin offered several pieces of evidence that favored evolution. In a subdued manner, he attempted to convince the scientific community of the validity of his theory.

Paleontology

One piece of evidence offered by Darwin is found in the science of paleontology. Paleontology deals with locating, cataloging, and interpreting the life forms that existed in past millennia. It is the study of fossils—the bones, shells, teeth, and other remains of organisms, or evidence of ancient organisms, that have survived over eons of time.

Paleontology supports the theory of evolution because it shows a descent of modern organisms from common ancestors. Paleontology indicates that fewer kinds of organisms existed in past eras, and the organisms were probably less complex. As paleontologists descend deeper and deeper into layers of rock, the variety and complexity of fossils decreases. The fossils from the uppermost rock layers are most like current forms. Fossils from the deeper layers are the ancestors of modern forms.

Comparative anatomy

More evidence for evolution is offered by comparative anatomy (Figure 12-1). As Darwin pointed out, the forelimbs of such animals as humans, whales, bats, and other creatures are strikingly similar, even though the forelimbs are used for different purposes (that is, lifting, swimming, and flying). Darwin proposed that similar forelimbs have similar origins, and he used this evidence to point to a common ancestor for modern forms. He suggested that various modifications are nothing more than adaptations to the special needs of modern organisms.

Figure 12-1 The forelimbs of a human and four animals showing the similarity in construction. This similarity was offered by Darwin as evidence that evolution has occurred.

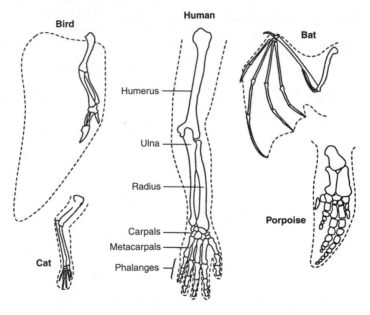

Darwin also observed that animals have structures they do not use. Often these structures degenerate and become undersized compared with similar organs in other organisms. The useless organs are called *vestigial organs.* In humans, they include the appendix, the fused tail vertebrae, the wisdom teeth, and muscles that move the ears and nose. Darwin maintained that vestigial organs may represent structures that have not quite disappeared. Perhaps an environmental change made the organ unnecessary for survival, and the organ gradually became nonfunctional and reduced in size. For example, the appendix in human ancestors may have been an organ for digesting certain foods, and the coccyx at the tip of the vertebral column may be the remnants of a tail possessed by an ancient ancestor.

Embryology

Darwin noted the striking similarity among embryos of complex animals such as humans, chickens, frogs, reptiles, and fish. He wrote that the uniformity is evidence for evolution. He pointed out that human embryos pass through a number of embryonic stages inherited from their ancestors because they have inherited the developmental mechanisms from a common ancestor. These mechanisms are modified in a way that is unique to an organism's way of life.

The similarities in comparative embryology are also evident in the early stages of development. For example, fish, bird, rabbit, and human embryos are similar in appearance in the early stages. They all have gill slits, a two-chambered heart, and a tail with muscles to move it. Later on, as the embryos grow and develop, they become less and less similar.

Comparative biochemistry

Although the biochemistry of organisms was not well known in Darwin's time, modern biochemistry indicates there is a biochemical similarity in all living things. For example, the same mechanisms for trapping and transforming energy and for building proteins from amino acids are nearly identical in almost all living systems. DNA and RNA are the mechanisms for inheritance and gene activity in all living organisms. The structure of the genetic code is almost identical in all living things. This uniformity in biochemical organization underlies the diversity of living things and points to evolutionary relationships.

Domestic breeding

From observing the domestic breeding experiments of animal and plant scientists, Darwin developed an idea about how evolution takes place. Domestic breeding brings about new forms that differ from ancestral stock. For example, pigeon fanciers have developed many races of pigeons through domestic breeding experiments. In effect, evolution has taken place under the guidance of human hands. The development of new agricultural crops by farmers and botanists provides more evidence for directed evolution.

Geographic distribution

Darwin was particularly interested in the life forms of the Galapagos Islands. He noticed how many of the birds and other animals on the islands were found only there. The finches were particularly puzzling because Darwin found 13 species of finches not found anywhere else in the world, as far as he knew. He concluded that the finches had evolved from a common ancestor that probably reached the island many generations earlier. In the isolation of the Galapagos Islands, the original finches had probably evolved into the 13 species.

Other geographic distributions also help to explain evolution. For instance, alligators are located only in certain regions of the world, presumably because they have evolved in those regions. The islands of Australia and New Zealand have populations of animals found nowhere else in the world because of their isolated environments.

Mechanisms of Evolution

Evolution does not occur in individuals but in populations. A **population** is an interbreeding group of individuals of one species in a given geographic area. A population evolves because the population contains the collection of genes called the gene pool. As changes in the gene pool occur, a population evolves.

Mutation

Mutation, a driving force of evolution, is a random change in a population's gene pool. It is a change in the nature of the DNA in one or more chromosomes. Mutations give rise to new alleles; therefore, they are the source of variation in a population.

Mutations may be harmful, but they may also be beneficial. For example, a mutation may permit organisms in a population to produce enzymes that will allow them to use certain food materials. Over time, these types of individuals survive, while those not having the mutations perish. Therefore, natural selection tends to remove the less-fit individuals, allowing more-fit individuals to survive and form a population of fit individuals.

Gene flow

Another mechanism of evolution may occur during the migration of individuals from one group to another. When the migrating individuals interbreed with the new population, they contribute their genes to the gene pool of the local population. This establishes **gene flow** in the population.

Gene flow occurs, for example, when wind carries seeds far beyond the bounds of the parent plant population. As another example, animals may be driven off from a herd. This forces them to migrate to a new population, thereby bringing new genes to a gene pool. Gene flow tends to increase the similarity between remaining populations of the same species because it makes gene pools more similar to one another.

Genetic drift

Another mechanism for evolution is genetic drift. **Genetic drift** occurs when a small group of individuals leaves a population and establishes a new one in a geographically isolated region. For example, when a small population of fish is placed in a lake, the fish population will evolve into one that is different from the original. Fitness of a population is not considered in genetic drift, nor does genetic drift occur in a very large population.

Natural selection

Clearly, the most important influence on evolution is natural selection, which occurs when an organism is subject to its environment. The fittest survive and contribute their genes to their offspring, producing a population that is better adapted to the environment. The genes of less-fit individuals are eventually lost. The important selective force in natural selection is the environment.

Environmental fitness may be expressed in several ways. For example, it may involve an individual's ability to avoid predators, it may imply a greater resistance to disease, it may enhance ability to obtain food, or it may mean resistance to drought. Fitness may also be measured as enhanced reproductive ability, such as in the ability to attract a mate. Better-adapted individuals produce relatively more offspring and pass on their genes more efficiently than less-adapted individuals.

Several types of natural selection appear to act in populations. One type, stabilizing selection, occurs when the environment continually eliminates individuals at extremes of a population. Another type of natural selection is disruptive selection. Here, the environment favors extreme types in a population at the expense of intermediate forms, thereby splitting the population into two or more populations. A third type of natural selection is directional selection. In this case, the environment acts for or against an extreme characteristic, and the likely result is the replacement of one gene group with another gene group. The development of antibiotic-resistant bacteria in the modern era is an example of directional selection.

Species development

A species is a group of individuals that share a number of features and are able to interbreed with one another. (When individuals of one species mate with individuals of a different species, any offspring are usually sterile.) A species is also defined as a population whose members share a common gene pool.

The evolution of a species is *speciation,* which can occur when a population is isolated by geographic barriers, such as occurred in the isolation of Australia, New Zealand, and the Galapagos Islands. The variety of life forms found in Australia but nowhere else is the characteristic result of speciation by geographic barriers.

Speciation can also occur when reproductive barriers develop. For example, when members of a population develop anatomical barriers that make mating with other members of the population difficult, a new species can develop. The timing of sexual activity is another example of a reproductive barrier. Spatial difference, such as one species inhabiting treetops while another species occurs at ground level, is another reason why species develop.

Gradual versus rapid change

Darwin's theory included the fact that evolutionary changes take place slowly. In many cases, the fossil record shows that a species changed gradually over time. The theory that evolution occurs gradually is known as *gradualism.*

In contrast to gradualism is the theory of *punctuated equilibrium,* which is a point of discussion among scientists. According to the theory of punctuated equilibrium, some species have long, stable periods of existence interrupted by relatively brief periods of rapid change.

Both groups of scientists agree that natural selection is the single most important factor in evolutionary changes in species. Whether the change is slow and gradual, or punctuated and rapid, one thing is certain: Organisms have evolved over time.

Chapter Checkout

Q&A

1. Each of the following techniques have been commonly used to vali-
date Darwin's theory of evolution, except _____.

   a. RFLP

   b. paleontology

   c. comparative anatomy

2. Wind carrying seeds to a new, uninhabited site is an example of
_____.

   a. gene flow

   b. genetic drift

   c. species development

3. A small group that leaves a population and establishes a new one in
a geographically isolated region is an example of _____.

   a. gene flow

   b. genetic drift

   c. speciation

Answers: 1. a **2.** a **3.** b

Chapter 13

THE ORIGIN AND EVOLUTION OF LIFE

Chapter Check-In

❑ Forming organic molecules

❑ Tracing the first eukaryotic cells

❑ Exploring the Paleozoic, Mesozoic, and Cenozoic Eras

Scientists hypothesize that the universe came into existence about 15 billion years ago with a colossal explosion often referred to as the *big bang*. The gases and dust from that explosion produced the earliest generation of stars, and over a period of billions of years, the stars exploded, and their debris formed other stars and planets. The solar system was presumably formed in this way 4 to 5 billion years ago. During the next billion years, the molten earth cooled, forming a hardened, outer crust. About 3.5 billion years ago, living things came into being.

Origin of Organic Molecules

About 3.8 billion years ago, earth's atmosphere consisted of such elements as nitrogen, hydrogen, sodium, sulfur, and carbon. Some of these elements combined to form hydrogen sulfide, methane, water, and ammonia. Water vapor in this mist probably caused millions of years of torrential rains, during which the oceans formed. Gas and water from the earth's core came to the surface through volcanoes. Ultraviolet radiation bathed the earth, and the elements and compounds interacted with one another to form complex molecules.

In 1953, Stanley Miller and Howard Urey performed a classic experiment in which they circulated methane, ammonia, water vapor, and hydrogen gas in a closed environment and passed electric sparks through it. After several days, they discovered that complex compounds of carbon had formed in the mixture. Their experiments indicated that in the primitive earth atmosphere, complex organic molecules could form, including amino acids, carbohydrates, and nucleic acids. The theory they expressed is the *primordial soup theory.*

Recent theories about the origin of organic molecules suggest that these molecules may have formed in hydrothermal vents deep in the oceans, where hot gases and elements emerge from cracks in the earth's crust. Living organisms have been found near these vents, lending credence to the theory.

Origin of Cells

The appearance of the first cells marked the origin of life on earth. However, before cells could form, the organic molecules must have united with one another to form more complex molecules called *polymers.* Examples of polymers are polysaccharides and proteins.

In the 1950s, Sidney Fox placed amino acids in primitive earth conditions and showed that amino acids would unite to form polymers called **proteinoids.** The proteinoids were apparently able to act as enzymes and catalyze organic reactions.

More recent evidence indicates that RNA molecules have the ability to direct the synthesis of new RNA molecules as well as DNA molecules. Because DNA provides the genetic code for protein synthesis, it is conceivable that DNA may have formed in the primitive earth environment as a consequence of RNA activity. Then DNA activity could have led to protein synthesis (see Chapter 10).

For a cell to come into being, some sort of enclosing membrane is required to hold together the organic materials of the cytoplasm. A generation ago, scientists believed that membranous droplets formed spontaneously. These membranous droplets called **protocells** were presumed to be the first cell. Modem scientists believe, however, that protocells do not carry any genetic information and lack the internal organization of cells. Thus the protocell theory is not widely accepted. Several groups of scientists are currently investigating the synthesis of polypeptides and short nucleic acids on the surface of clay. The first cells remain a mystery.

Ancient Life

Earth came into existence about 4.6 billion years ago, and about 3.8 billion years ago, the evolution of chemicals began. Scientists estimate that at about 3.5 billion years ago, the first cells were in existence.

Scientists believe that the first cells lived within the organic environment of the earth and used organic foods for their energy. The type of chemistry in those first cells was somewhat similar to **fermentation,** which uses organic molecules, such as glucose. The energy yield, although minimal, is enough to sustain living things (see Chapter 6). However, organic material would soon have been used up if this were the sole source of nutrition, so a new process had to develop.

The evolution of a pigment system that could capture energy from sunlight and store it in chemical bonds was an essential breakthrough in the evolution of living things. The organisms that possess these pigments are commonly referred to as **cyanobacteria** (at one time, they were called blue-green algae). These single-celled organisms produce carbohydrates by the process of photosynthesis. In doing so, they produce oxygen as a waste product (see Chapter 5). For a period of about 1 billion years, photosynthesis provided oxygen to the atmosphere, which gradually changed until it became oxygen rich, as it is today.

Another group of organisms that was present at the same time as the cyanobacteria was a group of bacteria called **archaebacteria.** Archaebacteria differ from "modern" bacteria (known as **eubacteria**) in that archaebacteria have a different ribosomal structure, different cell membrane composition, and different cell wall composition. The archaebacteria have been traced to a period of about 3 billion years ago. They are able to multiply at the very high temperatures that were present on the earth then, and their nutritional requirements reflect the composition of the primitive earth.

First Eukaryotes

The cyanobacteria and archaebacteria of the primitive earth are now referred to as **prokaryotes** (together with the modern bacteria). The prokaryotes are discussed in Chapter 16. Approximately 1.5 billion years ago, in an oxygen-containing atmosphere, the first **eukaryotes** came into being. Eukaryotes have a nucleus, a nuclear membrane, a number of organelles, a ribosomal structure different from that of prokaryotes, reproduction by mitosis, and other features that distinguish them from prokaryotes (see Chapter 3).

No one is certain how eukaryotes came into being. The *endosymbiotic theory* suggests that bacteria were engulfed by larger cells. The bacterial cells remained in the cell, assumed some of the chemical reactions for these cells, and became the mitochondria of these cells. The cells then reproduced and flourished. These cells became animal cells.

An extension of the endosymbiotic theory refers to plants. In this case, pigmented bacteria, such as the cyanobacteria, were engulfed by larger cells. The cyanobacteria remained in the cells and became the chloroplasts of these cells. Photosynthesis occurs in the chloroplasts of modern plant cells.

Life on Land

For billions of years, the only life present on earth existed in the nutrient environments of the oceans, lakes, and rivers. About 600 million years ago, the Paleozoic Era began. Scientists believe that living things first came to occupy the land during this era. They also believe that during a subdivision of the Paleozoic Era called the Cambrian Period, the main groups of marine invertebrates in existence today evolved. A so-called "Cambrian explosion" occurred. The appearance of multicellular organisms is notable in the Cambrian Period, when evolution and natural selection led to a vast array of organisms filling every conceivable niche on the earth. Many organisms that arose at that time have since become extinct.

After the Cambrian Period came the Ordovician Period. In the Ordovician Period, wormlike animals with stiff rods along their backs came into being. These organisms are now called **chordates,** which include reptiles, amphibians, birds, and mammals. A backbone composed of vertebrae was first developed during the next period, the Silurian Period. During the Devonian Period that followed, bony fish developed. The first terrestrial plants also evolved at about this time.

The next most recent era after the Paleozoic Era is the Mesozoic Era. The Mesozoic Era began about 250 million years ago. During this era, reptiles such as the dinosaurs evolved and became the predominant life form on earth. A mammal-like animal also evolved during this period. After the dinosaurs died out at the end of the Mesozoic Era, the mammal-like animal evolved into the modern **mammals.** Birds first appeared during the Mesozoic Era, and lush greenery spread over the earth.

During the Mesozoic Era, the continents existed as one landmass called Pangaea. Toward the end of the era, the landmass broke into smaller pieces, and the pieces moved apart to form the continents. The movement of the continents isolated plants and animals from one another, sparking the genetic drift that gives us different species of animals and plants in different parts of the world today.

At the end of the Mesozoic Era, a great extinction took place, and the dinosaurs disappeared. Although the reasons for the extinction are not clear, many scientists believe that a great meteor collided with earth, presumably in the Yucatan peninsula area of Mexico. The dust raised by this impact so darkened the earth that photosynthesis became impossible. The amount of available food became scarce, and the dinosaurs disappeared. Mammals filled the void left by the dinosaurs' absence.

The Cenozoic Era, the most recent era, has been called the age of mammals. The era began about 65 million years ago and extends to the present time. Mammals that survived the great extinctions at the end of the Mesozoic Era radiated throughout the earth and filled the niches once inhabited by the dinosaurs. Primates first appeared on the earth about 35 million years ago, and modern humans are believed to have evolved about 40,000 years ago.

Chapter Checkout

Q&A

1. Sidney Fox showed that amino acids could unite to form polymers called _____ that act as enzymes.

    a. protocells
    b. eukaryotes
    c. proteinoids

2. Organisms called _____ evolved a pigment system that could capture energy from sunlight to create energy.

    a. eukaryotes
    b. cyanobacteria
    c. archaebacteria

3. According to the endosymbiotic theory, bacterial cells ultimately became _____ within eukaryotic cells.

    a. nuclei
    b. mitochondria
    c. Golgi bodies

4. The _____ Era is know as the age of the mammals, whereas the _____ Era is when the reptiles, such as the dinosaurs, evolved to become the predominant life form on earth.

a. Cenozoic, Paleozoic
b. Paleozoic, Cenozoic
c. Cenozoic, Mesozoic

Answers: 1. c **2.** b **3.** b **4.** c

Chapter 14

HUMAN EVOLUTION

Chapter Check-In

❑ Walking upright

❑ Using tools

❑ Developing language

❑ Building societies

Fossils and fragments of jaws suggest to scientists that the ancestors of monkeys, apes, and humans began their evolution approximately 50 million years ago. Additional evidence for this evolution has been derived from experiments in biochemistry and changes that occur in the DNA of cells. Mutation rates in the DNA have been calculated, and evolutionary changes can be determined from these calculations. By noting mutation rates, scientists can estimate the time since two species diverged from a single ancestor, and they can develop a family tree.

Australopithecus

Scientific evidence indicates that the first hominids (humanlike creatures) belonged to a group called *Australopithecus*. Members of the genus Australopithecus are believed to have displayed a critical step in human evolution: the ability to walk upright on two feet.

In 1924, the complete skull of a young child was found in a limestone quarry in South Africa. The fossil was unlike any ever seen before. The fossil was named *Australopithecus africanus*. Since then, several other *A. africanus* fossils have been found, showing heavy jaws, low foreheads, and small brain capacities.

In 1974, a team led by Donald Johanssen and Tim White found a nearly complete skeleton of a female in the Afar region of Ethiopia. The fossil, which is earlier than A. africanus, is widely known as "Lucy," but it is classified as *Australopithecus afarensis*.

In 1977, Mary Leakey and her group made another important discovery: a set of fossil hominid footprints. The footprints, dated at some 4 million years of age, show that the two Australopithecenes that made the footprints walked erect.

Members of the Australopithecus genus are considered the first hominids but not the first humans. Their brains were small in comparison with human brains, and they had long, monkeylike arms. Other members of the Australopithecus genus have been classified as *A. robustus* and *A. boisei*. These different species of Australopithecus lived in Africa and are believed to have been primarily plant eaters. Members of the Australopithecus group eventually died out about 1 million years ago.

Homo Habilis

Fossils dating back about 2 million years have been found with brain capacities much larger than any Australopithecus fossil. On the basis of brain size, these fossils are named *Homo habilis*. Homo habilis is regarded as the first human and the first species of the genus *Homo*. Homo habilis means "handy human." Members of this species were apparently able to use tools, build shelters, and fashion protective clothing. Members of the species appear to have spent much time in trees as well as on the ground, and to have walked erect on two legs. Homo habilis is thought to have been the predominant species during the Stone Age, a time during which humans fashioned stone tools. Homo habilis eventually became extinct, presumably about 35,000 years ago.

Homo Erectus

The first hominid to leave Africa for Europe and Asia was *Homo erectus*. Evidence suggests that Homo erectus replaced Homo habilis. Homo erectus was about the size of modern humans and was fully adapted for upright walking. Its brain was much larger than its ancestor's brain, but it had features that separate it from modern humans. The tools of Homo erectus were more sophisticated than the tools of Homo habilis. Homo erectus was perhaps the first hunter-gatherer.

Homo erectus was skillful at hunting and butchering animals and is presumed to be the first fire user. Archaeologists believe Homo erectus established early cultures and had methods for communicating information to the next generation. The concept of language is believed to have existed in Homo erectus.

Homo Sapiens

The earliest fossils of *Homo sapiens* date to about 200,000 years ago. Homo sapiens means "intelligent human," and modern humans are classified in this species. Homo sapiens is believed to have evolved from Homo erectus. The evolution is thought to have taken place in Africa. The earliest fossils of Homo sapiens show a gradual change over the last 200,000 years into varieties of Homo sapiens, but not new species.

The oldest fossils classified as Homo sapiens are the Neanderthals, who lived from about 125,000 to about 35,000 years ago in Europe. Neanderthals were short, stocky, and powerfully built. They had large skulls and heavy faces with prominent brow ridges, a sloping forehead, and a heavy jaw but a small chin. Their large heads contained larger brains than those of today's humans. Neanderthals made tools of various types and lived in caves and huts.

While Neanderthals were still in existence, the oldest fully modern variety of Homo sapiens emerged. This modern variety is Cro-Magnon, for a cave in southwestern France where the first fossils were found. Cro-Magnon was somewhat similar to Neanderthals, but they had smaller heads and less prominent faces. Cro-Magnon is similar to modern humans. Cro-Magnon hunted, displayed culture, and showed the gradual development that led to today's societies. About 10,000 years ago, the first evidence of cities and social structure existed, and by about 5,000 years ago the first great civilizations began to flourish. Today's humans are the descendants of this variation of Homo sapiens.

Chapter Checkout

Q&A

1. The first humanlike creatures able to walk upright are called _____.

   a. Australopithecus
   b. Homo erectus
   c. Homo habilis
   d. Afarensis

2. _____ is considered to be the first human.

   a. Homo sapiens
   b. Homo habilis
   c. Homo erectus
   d. Homo afarensis

3. The first hunter-gatherer is which of the following?

   a. Homo erectus
   b. Homo habilis
   c. Homo sapiens
   d. Neanderthals

4. The concept of language first existed in which of the following populations?

   a. Homo habilis
   b. Homo erectus
   c. Homo sapiens
   d. Cro-Magnon

Answers: 1. a **2.** b **3.** a **4.** b

Chapter 15

THE UNITY AND
DIVERSITY OF LIFE

Chapter Check-In

❏ Outlining the biological classification system

❏ Understanding the five-kingdom classification scheme

The earth today is home to more than 300,000 species of plants and more than one million species of animals. Biologists called *taxonomists* have devised a carefully developed scheme to organize these myriad species. In the mid 1700s, Carolus Linnaeus, a Swedish physician and botanist, published several books in which he described thousands of plant and animal species. Linnaeus grouped the species according to their reproductive parts. Linnaeus developed the two-part binomial taxonomy system of categorizing organisms according to genus and species. Linnaeus's work remains valid. It has been combined with the work of Charles Darwin in the field of evolution to form the foundation of modern taxonomy. Darwin's theory of evolution states that all modern species are derived from earlier species and that all organisms, past and present, share a common ancestry. Darwin's theory of evolution, which has become a unifying theme in biology, is the organizing principle of modern taxonomy.

Taxonomists classify organisms in a way that reflects their biological ancestry. Because the ancestral relationships are complex, the taxonomic schemes are also complex. Despite their complexity, the taxonomic schemes provide considerable insight into the unity and diversity of life. The term "classification" is synonymous with the word "taxonomy."

Basics of Classification (Taxonomy)

All organisms in the living world are classified and named according to an international system of criteria that dates to the early part of this century.

The rules of classification establish a procedure to be followed when a new species is identified and named. (The rules of classification apply only to formal scientific names, not to common names.)

The scientific name of any organism, called the **binomial name,** has two elements. For example, humans have the binomial name *Homo sapiens.* The name of any species is two words: the name of the genus followed by the "species modifier." For humans, *Homo* is the genus and *sapiens* is the species modifier. The genus name is generally a noun, while the species modifier is an adjective. Thus, *Homo sapiens* means "human knowing."

The generally accepted criterion for defining a species is that organisms of the same species interbreed under natural conditions to yield fertile offspring. Individuals of different species normally do not mate. If they are forced to mate, the mating is either unsuccessful or the offspring are sterile. For example, a horse *(Equus caballus)* can be mated to a donkey *(Equus assinus),* and the result will be a mule. However, mules are sterile and cannot reproduce. Thus, the horse and donkey are classified as different species. A quarterhorse and a thoroughbred can mate and produce a fertile offspring. Therefore, both are classified as the same species: *Equus caballus.*

For humans, there is only one living species: Homo sapiens. However, in past ages other species, such as *Homo erectus,* may have coexisted with Homo sapiens. Homo erectus (see Chapter 14) is considered a separate species because presumably it could not mate with Homo sapiens.

The classification scheme provides a mechanism for bringing together various species into progressively larger groups. Taxonomists classify two species together in the same genus (the plural is *genera*). For example, the horse *Equus caballus* and the donkey *Equus assinus* are both placed in the genus *Equus.* Similar genera are brought together to form a **family.** Similar families are classified within an **order.** Orders with similar characteristics are grouped in a **class.** Related classes are grouped together as **divisions** or **phyla** (the singular is *phylum*). Divisions are used for plants and fungi, while phyla are used for animals and animal-like organisms. The largest and broadest category is the **kingdom.**

The classification of a human shows how the classification scheme works. Working from the top down, the human is classified first in the kingdom Animalia because it has the properties of animals. Animals are then divided into at least ten phyla, one of which is Chordata. Members of this phylum all have a backbone at some time in their lives.

Members of the phylum Chordata are then subdivided into various classes. Humans belong to the class Mammalia, together with other mammals (all of which possess mammary glands and nurse their young). The Mammalia are then divided into several orders, one of which is Primata. Humans belong to the order Primata along with other primates, such as gorillas and monkeys. The order Primata is subdivided into several families, one of which is Hominidae, the family that includes humans and humanlike creatures. Within the family of Hominidae is the genus *Homo,* which includes several species. One of these species is *Homo sapiens.*

Kingdoms of Living Things

In his classification scheme, Linnaeus recognized only two kingdoms of living things: Animalia and Plantae. At the time, microscopic organisms had not been studied in detail. They were placed either in a separate category called Chaos or, in some cases; they were classified with plants or animals. Then in the 1860s, the German investigator Ernst Haeckel proposed a three-kingdom system of classification. Haeckel's three kingdoms were Animalia, Plantae, and Protista. Members of the kingdom Protista included the protozoa, fungi, bacteria, and other microorganisms. Haeckel's system was not widely accepted, however, and microorganisms continued to be classified as plants (for example, bacteria and fungi) or animals (for example, protozoa).

Currently, the system of classification widely accepted by biologists is that devised by Robert Whittaker in 1968. Whittaker's classification scheme recognizes five kingdoms: Monera, Protista, Fungi, Plantae, and Animalia. The five-kingdom classification scheme is in general use today.

The kingdom Monera includes the bacteria and the cyanobacteria. These one-celled organisms are prokaryotic. Prokaryotic organisms have neither nucleus nor organelles in their cytoplasm, possess only a single chromosome, have small ribosomes, and reproduce by simple fission. Many of the organisms (called *autotrophic*) can synthesize their own foods, and some (called *heterotrophic*) digest preformed organic matter.

The second kingdom, Protista, includes the protozoa, the one-celled algae, and the slime molds. The cells of these organisms are eukaryotic. They are unicellular, and they may be autotrophic or heterotrophic. Eukaryotic organisms have a nucleus and organelles in their cytoplasm, possess multiple chromosomes, have large ribosomes, and reproduce by mitosis.

The third kingdom, Fungi, includes the yeasts, molds, mildews, mushrooms, and other similar organisms. The cells of this kingdom are eukaryotic and heterotrophic. Some fungal species are unicellular, whereas other species form long chains of cells and are called *filamentous* fungi. A cell wall containing chitin or cellulose is found in most members. Food is taken in by the absorption of small molecules from the external environment.

The fourth kingdom is Plantae. Classified here are the mosses, ferns, and seed-producing plants. All plant cells are eukaryotic and autotrophic. The organisms synthesize their own foods by photosynthesis, and their cell walls contain cellulose. All the organisms are multicellular.

The final kingdom, Animalia, includes animals. Animals without backbones (invertebrates) and with backbones (vertebrates) are included here. The cells are eukaryotic; the organisms are heterotrophic. All animals are multicellular, and none have cell walls. In the kingdom Animalia, biologists classify such organisms as sponges, hydras, worms, insects, starfish, reptiles, amphibians, birds, and mammals. The feeding form is one in which large molecules from the external environment are consumed, then broken down to usable parts in the animal body.

Chapter Checkout

Q&A

1. The ideas of _____ and _____ combined to form the foundation of modern taxonomy.
   a. Darwin, Linnaeus
   b. Darwin, Leakey
   c. Darwin, Mendel

2. The order of the classification system is which of the following?
   a. Kingdom, Phylum, Class, Order, Family, Genus, Species
   b. Kingdom, Phylum, Order, Class, Family, Genus, Species
   c. Kingdom, Phylum, Class, Family, Order, Genus, Species

3. The kingdom _____ contains the bacteria and prokaryotic organisms whereas the kingdom _____ includes the algae and protozoa.
   a. Protista, Fungi
   b. Protista, Monera
   c. Monera, Protista

Answers: 1. a **2.** a **3.** c

Chapter 16
MONERA

Chapter Check-In

❏ Understanding the functions of bacteria

❏ Recognizing the role of cyanobacteria

❏ Comparing bacteria and the viruses

Members of the kingdom Monera are microscopic organisms that include the bacteria and cyanobacteria. Both the bacteria and cyanobacteria are prokaryotes. **Prokaryotes** lack a nucleus, and they have no organelles except ribosomes. The hereditary material exists as a single loop of double-stranded DNA in a nuclear region, or nucleoid. Members of the kingdom multiply by an asexual process called *binary fission*. No evidence of mitosis is apparent in the reproductive process.

Bacteria

Bacteria live in virtually all the environments on earth, including the soil, water, and air. They have existed for approximately 3 billion years, and they have evolved into every conceivable ecological niche on, above, and below the surface of the earth.

Most bacteria can be divided into three groups according to their shapes. The spherical bacteria are referred to as cocci (the singular is coccus); the rod-shaped bacteria are **bacilli** (the singular is *bacillus*); and the spiral bacteria are spirochetes if they are rigid or **spirilla** (the singular is *spirillum*) if they are flexible.

Cocci may occur in different forms. Those cocci that appear as an irregular cluster are *staphylococci* (the singular is *staphylococcus*) and are the cause of "staph" infections. Cocci in beadlike chains are *streptococci,* and bacteria in pairs are *diplococci.* One streptococcus is the cause of "strep throat," while another is a harmless organism used to make yogurt. A species of diplococcus is the agent of pneumonia, while a second causes gonorrhea, and a third is the agent of meningitis.

Characteristics of bacteria

Most bacterial species are **heterotrophic,** that is, they acquire their food from organic matter. The largest number of bacteria are *saprobic,* meaning that they feed on dead or decaying organic matter. A few bacterial species are *parasitic.* These bacteria live within host organisms and cause disease.

Certain bacteria are **autotrophic,** that is, they synthesize their own foods. Such bacteria often engage in the process of photosynthesis. They use pigments dissolved in their cytoplasm for the photosynthetic reactions (see Chapter 5). Two groups of *photosynthetic* bacteria are the green sulfur bacteria and the purple bacteria. The pigments in these bacteria resemble plant pigments. Some autotrophic bacteria are *chemosynthetic.* These bacteria use chemical reactions as a source of energy and synthesize their own foods using this energy.

Bacteria may live at a variety of temperatures. Bacteria living at very cold temperatures are *psychrophilic,* while those species living at human body temperatures are said to be *mesophilic.* Bacteria living at very high temperatures are *thermophilic.* Bacteria that require oxygen for their metabolism are referred to as aerobic, while species that thrive in an oxygen-free environment are said to be anaerobic. Some bacteria can live with or without air; they are described as *facultative.* Most bacterial species live in a neutral pH environment (about pH 7), but some bacteria can live in acidic environments (such as in yogurt and sour cream) and others can live in alkaline environments. Certain bacteria are known to live at the pH of 2 found in the human stomach.

Activities of bacteria

Bacteria play many beneficial roles in the environment. For example, some species of bacteria live on the roots of pod-bearing plants (legumes) and "fix" nitrogen from the air into organic compounds that are then available to plants. The plants use the nitrogen compounds to make amino acids and proteins, providing them to the animals that consume them. Other bacteria are responsible for the decay that occurs in landfills and the other debris in the environment. These bacteria recycle the essential elements in the organic matter.

In the food industry, bacteria are used to prepare many products, such as cheeses, fermented dairy products, sauerkraut, and pickles. In other industries, bacteria are used to produce antibiotics, chemicals, dyes, numerous vitamins and enzymes, and a number of insecticides. Today, they are used in genetic engineering to synthesize certain pharmaceutical products that cannot be produced otherwise (see Chapter 11).

In the human intestine, bacteria synthesize several vitamins not widely obtained in food, especially vitamin K. Bacteria also often break down certain foods that otherwise escape digestion in the body.

Unfortunately, many bacteria are **pathogenic;** that is, they cause human disease. Such diseases as tuberculosis, gonorrhea, syphilis, scarlet fever, food poisoning, Lyme disease, plague, tetanus, typhoid fever, and most pneumonias are due to bacteria. In many cases, the bacteria produce powerful toxins that interfere with normal body functions and bring about disease. The botulism (food poisoning) and tetanus toxins are examples. In other cases, bacteria grow aggressively in the tissues (for example, tuberculosis and typhoid fever), destroying them and thereby causing disease.

Other bacteria

There are some exceptionally small forms of bacteria that can be seen only with the electron microscope. The rickettsiae are one group. These ultramicroscopic organisms are usually transmitted by arthropods such as fleas, ticks, and lice. They cause several human diseases including Rocky Mountain spotted fever and typhus.

Another type of ultramicroscopic bacteria is the chlamydiae. Like the rickettsiae, the chlamydiae can be seen only with the electron microscope. In humans, chlamydiae cause several diseases, including an eye infection called trachoma and a sexually transmitted disease called chlamydia.

Probably the smallest forms of bacteria are the mycoplasmas. Mycoplasmas occur in many shapes and have no cell walls. This latter characteristic distinguishes them from the other bacteria (where cell walls are prominent). Mycoplasmas can cause a type of pneumonia.

Another type of bacteria is the **archaebacteria.** Archaebacteria are ancient species of bacteria identified in recent years. They are separated from the other bacteria on the basis of their ribosomal structure and metabolic patterns. Archaebacteria are anaerobic species that use methane production as a key step in their energy metabolism. They are found in marshes and swamps. Some scientists believe there are two major subdivisions of bacteria: the Archaebacteria and all others, which they designate Eubacteria.

Cyanobacteria

Cyanobacteria are those organisms formerly known as **blue-green algae.** These members of the Monera kingdom are photosynthetic. Most are found in the soil and in freshwater and saltwater environments. The majority of species are unicellular, but some may form filaments. Like the other bacteria, all cyanobacteria are prokaryotes.

Cyanobacteria, which are autotrophic, serve as important fixers of nitrogen in food chains. In addition, cyanobacteria, a key component of the plankton found in the oceans and seas, produce a major share of the oxygen present in the atmosphere, while also serving as food for fish. Some species of cyanobacteria coexist with fungi to form **lichens.**

Cyanobacteria have played an important role in the development of the earth. Scientists believe that they were among the first photosynthetic organisms to occur on the earth's surface. Beginning about 2 billion years ago, the oxygen produced by cyanobacteria enriched the earth's atmosphere and converted it to its modern form. This conversion made possible all life-forms that use oxygen in their metabolisms.

Viruses

Technically, viruses are not members of the Monera kingdom. They are considered here because, like the bacteria, they are microscopic and cause human diseases. **Viruses** are acellular particles that lack the properties of living things but have the ability to replicate inside living cells. They have no energy metabolism, they do not grow, they produce no waste products, they do not respond to stimuli, and they do not reproduce independently. In the view of biologists, they are probably not alive.

Viruses consist of a central core of either DNA or RNA surrounded by a coating of protein. The core of the virus that contains the genes is the **genome,** while the protein coating is the *capsid.* Viruses have characteristic shapes. Certain viruses have the shape of an *icosahedron,* a 20-sided figure made up of equilateral triangles. Other viruses have the shape of a *helix,* a coil-like structure. The viruses that cause herpes simplex, infectious mononucleosis, and chickenpox are icosahedral. The viruses that cause rabies, measles, and influenza are helical.

Viruses reproduce only within living cells. They attach to the plasma membrane of the host cell and release their nucleic acid into the cytoplasm of the cell. The capsid may remain outside the cell, or it may be digested by the host cell within the cytoplasm. In the host cytoplasm, the DNA or RNA of the viral genome encodes the proteins that act as enzymes for the synthesis of new viruses. The enzymes use amino acids in the cell for protein synthesis and nucleotides from the host DNA for nucleic acid synthesis. The viruses obtain cellular ATP and use cellular ribosomes for additional viral synthesis. After some minutes or hours, the new viral capsids and genomes combine to form new viruses.

Once formed, the viruses may escape the host cell when the host cell disintegrates. Alternately, the new viruses may force their way through the plasma membrane of the cell and assume a portion of the plasma membrane as a viral envelope. In either process, the cell is often destroyed and hundreds of new viruses are produced.

Viruses can cause a number of human diseases, including measles, mumps, chickenpox, AIDS, influenza, hepatitis, polio, and encephalitis. Protection from these diseases can be rendered by using vaccines composed of weak or inactive viruses. A viral vaccine induces the immune system to produce antibodies, which provide long-term protection against a viral disease.

Chapter Checkout

Q&A

1. Each of the following are groups of bacteria, except _____.
    a. cocci
    b. spirochetes
    c. bacilli
    d. cyanobacteria

2. Bacteria that are able to live both with or without air are called _____.
    a. thermophilic
    b. facultative
    c. aerobic
    d. anaerobic

3. True or False: Prokaryotes contain the hereditary material in the nucleus of the cell.

4. Viruses are known to cause each of the following diseases except
_____.

    a. chlamydia
    b. influenza
    c. rabies
    d. infectious mononucleosis

5. True or False: Viruses are members of the kingdom Monera.

Answers: 1. d **2.** b **3.** False **4.** a **5.** False

Chapter 17
PROTISTA

Members of the kingdom Protista are a highly varied group of organisms, all of which are eukaryotic. In addition, protists are unicellular or, in some cases, colonial. Many species are autotrophs, creating their own food, while others are heterotrophs, feeding on organic matter. Many species are nonmotile, but the majority of protists are able to move by various means. Many protists have contractile vacuoles, which help them to remove excessive amounts of water from their cytoplasm. The kingdom Protista includes the protozoa, the slime molds, and the simple algae.

Protozoa

The protozoa are subdivided into four phyla depending on their method of locomotion: Mastigophora, Sarcodina, Ciliophora, and Sporozoa.

Mastigophora

Members of the phylum Mastigophora move about by using one or more whiplike flagella. The genus *Euglena* contains flagellated species. Members are freshwater protists with typical eukaryotic properties, including two flagella, reproduction by mitosis, and flexible nutritional requirements. *Euglena* species also possess chlorophyll within chloroplasts. This pigment allows the organisms to synthesize organic compounds in the presence of sunlight. When no sunlight is available, the organism feeds on dead organic matter in the surrounding environment. Thus, the organism is autotrophic and heterotrophic. Some biologists consider *Euglena* to be the basic stock of evolution for both animals and plants.

Certain species of Mastigophora are zooflagellates, while some are phytoflagellates. The *zooflagellates* live within the bodies of animals and are typified by the wood-digesting flagellates in the intestines of termites. Among the pathogenic zooflagellates are those that cause sleeping sickness, trichomoniasis, and giardiasis. The *phytoflagellates* have photosynthetic abilities and are often discussed with algae in textbooks.

Some species of Mastigophora organize themselves into colonies. Members of the genus *Volvox* are typical colonial forms. The cell colonies are not differentiated into tissues or organs, but the colonies show how a preliminary step in evolutionary development might have occurred.

Sarcodina

Members of the phylum Sarcodina are the **amoebas** and their relatives. Amoebas consist of a single cell without a definite shape. They feed on small organisms and particles of organic matter, and they engulf the particles by phagocytosis. Extensions of the cytoplasm called *pseudopodia* (the singular is *pseudopodium*) assist phagocytosis and motion in the organisms.

Amoebas are found in most lakes, ponds, and other bodies of fresh water. They move by a creeping form of locomotion called amoeboid motion. One amoeba called *Entamoeba histolytica* causes a type of dysentery in humans.

Two interesting amoebas are the foraminiferans and the radiolarians. Both are marine amoebas that secrete shells. Their shells have been identified as markers for oil deposits because both were present in the ocean communities that became the organic deposits that, under pressure, became oil fields.

Ciliophora

Members of the phylum Ciliophora move by means of cilia. The organisms are all heterotrophic and have specialized organelles in their cytoplasm. For example, they have two nuclei: a large macronucleus and a number of smaller micronuclei. The micronuclei carry the genetic information of the cell.

The ciliate *Paramecium* typifies the phylum Ciliophora. This organism has a slipper-shaped body with a covering called a *pellicle*. Defensive organelles called *trichocysts* are present in the pellicle. The organism reproduces by mitosis and by an elaborate form of sexual behavior called *conjugation*, which occurs when two *Paramecium* join to one another in the oral region and exchange nuclear material. The cilia of *Paramecium* provide a precise form of motion not provided by flagella or pseudopodia. The cilia can propel the *Paramecium* either forward or backward and move it in a spiral manner.

Sporozoa

Members of the phylum Sporozoa are exclusively parasites. They are so named because some members produce sporelike bodies. Often they have an amoeboid body form, but they are not related to the Sarcodina.

Sporozoans are generally parasitic organisms with complex life cycles involving several stages. One of the best-known members of the group is the *Plasmodium* species, which are the agents of malaria. The organisms spend portions of their life cycle within mosquitoes. After being injected into the human bloodstream by the mosquito, the parasites invade the red blood cells, undergo numerous changes, and emerge from the red blood cells, destroying them. The infected human experiences a malaria attack soon after.

Another important member of the Sporozoa group is *Toxoplasma gondii,* which causes toxoplasmosis, a disease of the white blood cells. Toxoplasmosis is normally not a dangerous disease, but pregnant women can pass it to their offspring, where it can cause tissue damage. Also, persons with AIDS are susceptible to severe cases of toxoplasmosis.

Two other important pathogens of the Sporozoa group are *Cryptosporidium coccidi* and *Pneumocystis carinii.* Both of these organisms cause severe disease in AIDS patients. The first causes a severe intestinal diarrhea that is accompanied by the loss of substantial volumes of water; the second causes a type of pneumonia. Both of these diseases typify the opportunistic diseases that occur when a person is infected with HIV.

Slime Molds

Although the **slime molds** have properties that resemble fungi, many scientists classify the organisms with the protists because of their protozoalike qualities. Slime molds may be "true" slime molds, which consist of a single, flat, very large cell with many nuclei, or they may be "cellular" slime molds, amoebalike cells that live independently and unite with other cellular slime molds to form a single, large, flat cell with many nuclei.

Though scientists are uncertain how slime molds evolved, the organisms show a hint of cellular cooperation that is characteristic of more complex, multicellular organisms. Cellular slime molds normally move about like amoebas. The organisms exist as a mass of cytoplasm with **diploid nuclei.** This mass is a *plasmodium.* Cytoplasmic streaming can be seen within the plasmodium. Slime molds are excellent research tools because they are large and easy to cultivate.

As long as a food supply is adequate and other environmental conditions are optimal, a slime mold grows indefinitely in its plasmodial stage. However, if conditions become harsh, the plasmodium can join with other plasmodia to form a large sluglike mass. This giant plasmodium can transform itself into a spore-bearing structure similar to a fungus. A stalk arises from the plasmodium, and nuclei within the stalk cells divide by meiosis. Knobs called *sporangia* develop at the end of the stalk and fill with haploid spores. The spores are shed and removed by the wind when the stalk dries. Eventually, the spores germinate into flagellated swarm cells. Swarm cells function as gametes and fuse to form a diploid zygote, which divides by mitosis and forms a new plasmodium. The plasmodium represents the new generation of slime mold.

Simple Algae

The term **algae** is not a formal biological term. It refers to a large number of photosynthetic organisms that are generally unicellular, rather simple, and not classified with plants. The organisms are plantlike, however, because they contain chloroplasts with chlorophyll. Most algae can be found in the oceans, but freshwater forms are also abundant.

The simple algae are subdivided into several divisions (rather than phyla, like the protozoa): Rhodophyta, Pyrophyta, Chrysophyta, Phaeophyta, and Chlorophyta. The divisions are based in part on the types of pigments and colors they have.

Rhodophyta

Rhodophyta is the division of red algae. These organisms are almost exclusively marine types. Most are unicellular, but some multicellular forms grow anchored to rocks below the level of the low tide. Some are large enough to be seaweeds. Red algae carry on photosynthesis using chlorophyll *a*. The red pigments are very similar to those in many species of cyanobacteria. A derivative of red algae called **agar** is commonly used in bacteriological media in the laboratory.

Pyrophyta

Members of the Pyrophyta are dinoflagellates. *Dinoflagellates* are unicellular organisms that are usually surrounded by thick plates that give them an armored appearance. Two flagella move the organism. Many dinoflagellates are luminescent. When affected by sudden movements, they give off light. When optimal conditions exist in the oceans, the dinoflagellates reproduce at explosive rates. Their red pigments cause the water to turn the color of blood. This condition is the *red tide*.

Chrysophyta

Members of the division Chrysophyta are golden algae, most of which are diatoms. *Diatoms* have cell walls or shells composed of two overlapping halves impregnated with silica. In the oceans, the diatoms carry on photosynthesis. They serve as an important source of food in the oceanic food chains. Diatomaceous earth, a light-colored porous rock composed of the shells of diatoms, is made into a commercial product called diatomite. Diatomite is used as a filler, absorbent, and filtering agent.

Phaeophyta

Members of the division Phaeophyta are the brown algae. These organisms, which are multicellular, are found almost exclusively in salt water, where they are known as rock weeds and kelps. Despite their great size, the tissue organization in these algae is quite simple compared with the other plants. Often they are used as fertilizers and sources of iodine.

Chlorophyta

Members of the division Chlorophyta are green algae, some of which are flagellated. The common laboratory specimen *Spirogyra* is classified here, as are other flagellated species. Many colonial forms are also classified as Chlorophyta; the colonial forms may represent the first evidence of multicellularity in evolution. *Ulva,* the common sea lettuce, is classified here.

Chapter Checkout

Q&A

1. The protozoa are classified into each of the following phyla, except
_____.

   a. Ciliophora
   b. Pyrophyta
   c. Sporozoa

2. The amoebas are members of the _____ phylum whereas the Paramecium typifies the phylum _____.

   a. Sporozoa, Ciliophora
   b. Mastigophora, Sarcodina
   c. Sarcodina, Ciliophora

3. True or False: Members of the phylum Sporozoa are exclusively parasites.

Answers: 1. b **2.** c **3.** True

Chapter 18
FUNGI

Chapter Check-In

❑ Classing up the kingdom Fungi
❑ Tracing sources of fungal disease
❑ Introducing the lichens

Fungi, together with bacteria and protists, are the major decomposers of organic matter on earth. Most fungi are *saprobes,* that is, they digest nonliving organic matter, such as wood, leaves, and dead animals. However, some fungi are parasites that attack living things and cause disease. Fungi cause many agricultural diseases as well as several human diseases.

A unique physical structure and the method by which they obtain nutrients distinguish fungi from the other four kingdoms. Fungi secrete enzymes into the environment and break down organic matter, and then absorb the small nutrient particles through their cell membranes. This process is called *extracellular digestion.*

Fungi are considered by many biologists to be multicellular organisms, and the body of a fungus consists primarily of cells joined in filaments. Each microscopic filament of a fungus is a *hypha* (the plural is *hyphae*). Hyphae may form a huge tangled interwoven network called a *mycelium* (the plural is *mycelia*). A mycelium is a visible structure. Fungi cells are unique because they possess a polysaccharide called *chitin*. In some cases, the cell walls also contain cellulose. Fungi live in environments that are generally acidic, and they prefer carbohydrate-rich foods.

Reproduction in fungi generally occurs via an asexual process. In some cases, filaments break from the main mycelium and grow into new individuals. Alternately, a fungus may produce spores by an asexual process. The spores disperse, germinate, divide, and produce genetically identical fungi. The hyphae and spores are usually haploid (contain one set of chromosomes), and the haploid phase dominates the life cycle of a fungus. Spores can withstand extreme dryness and cold to produce a new fungus when conditions permit.

Fungi can also reproduce by a sexual process, which results in a short-lived diploid cell (with two sets of chromosomes) that soon produces haploid cells through meiosis. The spores develop into cells that divide by mitosis to form a new hypha, and then a new mycelium.

The members of the kingdom Fungi are classified in different ways. Some biologists include the slime molds in the kingdom Fungi rather than with the Protista. In the most accepted classification scheme, the true fungi are placed in the phylum/division Eumycota and are subdivided into five major classes: Oomycetes, Zygomycetes, Ascomycetes, Basidiomycetes, and Deuteromycetes.

Oomycetes

Fungi of the class Oomycetes are generally *water molds,* a reference to the fact that most species are aquatic. During sexual reproduction, the members of this class form clusters of egglike bodies at the tips of their hyphae. Nearby hyphae grow toward the bodies and fuse with them. Nuclear fusions lead to the formation of sexual spores called oospores that germinate to produce new hyphae.

In the sexual process of reproduction, oomycetes form a unique cell called a zoospore. The zoospore has flagella and is able to move like an animal cell. Certain oomycetes cause downy mildew of grapes, white rust of cabbage, and the late blight of potatoes. Aquatic oomycetes infect fish in aquaria and nature.

Zygomycetes

A second class of fungi, Zygomycetes, includes mostly terrestrial fungi. Because the hyphae have no cross walls between the cells, they are said to be *coenocytic.* Sexual reproduction in these organisms occurs when sexually opposite hyphae fuse and form spores called *zygospores.*

A common member of this class is the bread mold *Rhizopus stolonifer.* This fungus forms a white or gray mycelium on bread. The sporangia containing the asexual spores can be seen extending into the air. One species of *Rhizopus* is used to ferment rice to sake, and another species is used in the production of cortisone.

Ascomycetes

Members of the class Ascomycetes are diverse. They range from unicellular yeasts to powdery mildews, cottony molds, and large complex "cup" fungi. In the latter, the hyphae are tightly packed and form a cup-shaped structure.

During sexual reproduction, the ascomycetes form a sac called an *ascus.* The ascus forms where sexually opposite hyphae have fused. Within the sac a number of ascospores form, and each can reproduce the entire fungus.

Within the Ascomycetes class is the yeast *Saccharomyces.* This organism is used in fermentation processes and baking. The producer of penicillin, *Penicillium,* is also in this class, as is *Aspergillus* a producer of citric acid, soy sauce, and vinegar. The chestnut blight and Dutch elm diseases are caused by ascomycetes.

Basidiomycetes

Members of the class Basidiomycetes are known as *club fungi.* They include the common mushroom, the shelf fungi, puffballs, and other fleshy fungi. Sexual spores called *basidiospores* are formed on clublike structures called *basidia* (the singular is *basidium*).

Perhaps the most familiar member of this class is the edible mushroom. The mycelium forms below ground. After the hyphae have fused, a mushroom cap emerges. Basidia form on the cap's underside along the gills, and basidiospores form on the basidia. Some mushrooms are edible, while some are poisonous, and they are similar in their form and shape.

Basidiomycetes also cause agricultural diseases, including rust and smut diseases. These diseases affect corn, blackberries, and a number of grains such as wheat, oats, and rye.

Deuteromycetes

Those fungi that lack a known sexual reproduction cycle are classed as Deuteromycetes. These fungi reproduce only by an asexual process, so far as is known. (When such an organism's sexual stage is discovered, the fungus is usually reclassified.)

Many human pathogens are currently classified as deuteromycetes. These organisms reproduce by fragmenting, with hyphae segments commonly blown about in dust by currents and the wind. A familiar deuteromycete is the fungus of athlete's foot, which can be picked up from fragments on towels and shower room floors.

Fungal Disease

Several fungi cause diseases in humans, some of them serious. For example, a disease of the lungs and spinal cord is caused by a fungus called *Cryptococcus neoformans*. This disease is serious in AIDS patients, and it is often spread in dust by currents.

Another human pathogen is *Candida albicans*. This organism causes disease of the oral cavity (thrush), as well as "yeast disease" of the reproductive tract. Normally the disease is mild, but in persons with HIV infection, it can be serious.

Other human fungal diseases include ringworm and athlete's foot. Each is caused by fungi of various genera, and each is characterized by blisterlike regions on the skin or in the webs of toes or fingers. Fungal diseases of the lung tissues include histoplasmosis, blastomycosis, and coccidioidomycosis.

Lichens

Lichens are associations between fungi and cyanobacteria (formerly called blue-green algae). The cyanobacteria are the photosynthetic elements in the association, which is an example of **mutualism** (a living arrangement where both partners benefit). The photosynthetic organisms provide nutrients for themselves and the fungus, while the fungus provides water and minerals for the photosynthetic metabolism.

Lichens exist in harsh habitats with extreme climatic conditions. They are found on mountaintops, on rock faces in the desert, and on tree bark. Commonly, lichens set the stage for the growth of more complex plants. Lichens can recover from freezing and drying to resume normal metabolic activities. Their growth is so slow that some patches of lichens may be thousands of years old.

Chapter Checkout

Q&A

1. Each of the following terms is a class of fungi, except _____.

   a. Zygomycetes
   b. Ascomycetes
   c. Deuteromycetes
   d. Lichens

2. The producer of penicillin, *Penicillium*, is a member of which of the following classes of Fungi?

   a. Zygomycetes
   b. Ascomycetes
   c. Deuteromycetes
   d. Basidiomycetes

3. The water molds are examples of organisms of which class of Fungi?

   a. Oomycetes
   b. Zygomycetes
   c. Ascomycetes
   d. Basidiomycetes

4. Lichens are associations between fungi and the blue-green algae in which both partners benefit. This arrangement is commonly known as _____.

   a. mutualism
   b. mycelium
   c. chitin
   d. basidia

Answers: 1. d **2.** b **3.** a **4.** a

Chapter 19

PLANTS: DIVERSITY AND REPRODUCTION

Chapter Check-In

❑ Distinguishing characteristics of plants

❑ Introducing the nonvascular plants

❑ Understanding the vascular architecture

Plants are multicellular eukaryotic organisms with the ability to produce their own food by the process of photosynthesis. (They are autotrophs.) Algae have historically been included with the plants, but they are now classified with the protists. The modern definition of plants includes organisms that live primarily on land (and sometimes in water), excluding algae that live primarily in water.

Another distinguishing characteristic of plants is their type of chlorophyll. **Chlorophyll** is used to absorb energy from the sun during the process of photosynthesis. Plants have chlorophyll *a* and chlorophyll *b,* while many species of algae do not have chlorophyll *b.* Green algae have both types of chlorophyll, but they are single-celled forms. Many evolutionary biologists believe that the green algae gave rise to the land plants.

Plants occur in two major groups: the nonvascular plants and the vascular plants. **Nonvascular plants** do not have specialized tissues to transport fluids, while **vascular plants** do have specialized tissues. The bryophytes (the mosses and liverworts) are the only major group of nonvascular plants. There are three large groups of vascular plants: the seedless vascular plants (for example, ferns), the vascular plants with naked seeds (for example, pines), and the vascular plants with protected seeds (for example, flowering plants). While animals are classified in phyla, the plants are classified in divisions.

The life cycle of plants has both a multicellular haploid and multicellular diploid phase. Because both phases of the life cycle are multicellular, this type of life cycle is an *alternation of generations*. In contrast, animal life cycles have a multicellular diploid phase and a unicellular haploid phase.

The alternating generations of plants are the sporophyte generation and the gametophyte generation. Individuals in the gametophyte generation (often called *gametophytes*) form gametes, or sex cells. Gametes are haploid cells (they contain one set of chromosomes). Haploid gametes fuse in fertilization, a process that takes place in water. This fusion produces fertilized eggs, which are diploid cells (they have two sets of chromosomes). The plants that develop are diploid plants of the sporophyte generation. Individuals in the sporophyte generation (*sporophytes*) undergo meiosis to produce haploid spores.

Plants produce their gametes in specialized structures. In the nonvascular bryophytes and in the vascular plants, the egg cells are formed in structures called *archegonia* (the singular is *archegonium*). Sperm cells are produced in structures called *antheridia* (the singular is *antheridium*). In some specialized plants, these structures are reduced, and the sporophyte generation is dominant over the gametophyte generation in the life cycle.

Nonvascular Plants

Nonvascular plants belong to the division Bryophyta, which includes mosses, liverworts, and hornworts. These plants have no vascular tissue, so the plants cannot retain water or deliver it to other parts of the plant body. The bryophytes do not possess true roots, stems, or leaves, although the plant body is differentiated into leaflike and stemlike parts. In some species, there are rootlike structures called rhizoids. With no vascular tissue, the bryophytes cannot retain water for long periods of time. Consequently, water must be absorbed directly from the surrounding air or another nearby source. This explains the presence of mosses in moist areas, such as swamps and bogs, and on the shaded sides of trees.

The life cycle of the moss is typical of the bryophytes. Flask-shaped archegonia, located among the top leaves of the female gametophytes, produce one egg cell each. Antheridia, located similarly on the male gametophyte, produce many sperm cells that swim in drops of rainwater or dew into the neck of the archegonium to fertilize the egg cell.

The zygote that results from the fertilization develops into a young sporophyte within the archegonium. The sporophyte grows out of the archegonium, taking its nourishment from the gametophyte, and differentiates into a slender stalk with a spore capsule near the tip. Haploid spores are produced by meiosis in this capsule, and when the tip of the capsule opens, the spores are freed. The spores settle in the soil and germinate into gametophytes, which represent the next stage in the alternation of generations.

The life cycles of all bryophytes are uniform, and although the gametophyte generation is the most familiar aspect of the life cycle, neither the sporophyte nor gametophyte generation is dominant.

Vascular Plants

The vascular plants encompass several divisions of plants and are collectively known as tracheophytes. **Tracheophytes** are characterized primarily by the presence of a vascular system composed of two types of specialized tissue: xylem and phloem. **Xylem** conducts water and minerals upward from the roots of a plant, while **phloem** transports sugars and other nutrients from the leaves to the other parts of the plant. Both xylem and phloem are distributed throughout the plant. The vascular tissue also serves as a means for mechanical support in the plant, so some tracheophytes (such as trees) can grow quite tall.

Seedless vascular plants (ferns)

In many plants, seeds are the structures from which the sporophyte generation emerges. Seeds protect the embryonic plant during its early stages and store food. Many plants do not form seeds in their life cycles, but they have flourished nevertheless. Among the seedless vascular plants are the ferns, classified in the division Pteridophyta.

A mature fern plant produces spores by meiosis. The spores are stored in cases called *sori* (the singular is *sorus*) on the underside of the fern leaf. The leaf is a *frond*, and the sori resemble dots on the underside of the frond. The fern plant is the sporophyte generation, and this generation dominates the life cycle.

After the spores are dispersed, they germinate into small heart-shaped haploid plants if they reach moist ground. Each plant is a *prothallus*. These plants are the gametophytes. The antheridia and archegonia are found on the underside of the prothallus. Sperm produced in the antheridia swim through moisture on the plant to the archegonia and fertilize the egg cells. The zygote that results grows within the protection of the archegonium and develops into a young sporophyte. The sporophyte eventually grows out of the archegonium into a slender stalk. The stalk develops into a sporophyte, which in its adult form is the familiar fern plant.

Other divisions of the seedless vascular plants are the Psilophyta (the whisk ferns), the Lycophyta (the club mosses), and the Sphenophyta (the horsetails).

Vascular plants with unprotected seeds (gymnosperms)

The vascular plants having naked seeds are known as gymnosperms. Their seeds are not enclosed in female tissues and are therefore said to be naked. There are four divisions of gymnosperms: Cycadophyta (Cycads), Ginkgophyta (Ginkgo), Gnetophyta (Gnetae), and Coniferophyta (Conifer).

Coniferophyta is the largest and most familiar division of the gymnosperms. These plants are cone-bearers and are therefore called *conifers.* The seeds are borne on the surface of the female cone scales. Members of this division include trees such as cedars, firs, spruces, pines, and giant redwoods. The leaves are generally needle-shaped and contain vascular tissue.

The full-grown conifer (for example, a pine tree) is the sporophyte generation of the plant. The sporophyte produces male and female cones on the same tree. These cones produce spores that undergo meiosis and produce the male and female gametophytes. Male gametophytes are the *pollen grains,* each consisting of four cells. The male gametophyte produces sperm cells in the pollen grains. The female gametophyte produces two or three egg cells that develop within protective structures called *ovules.*

In the spring, the male cone releases pollen, which is blown about by the wind. Some pollen gets trapped on the female cone where it germinates and forms a pollen tube that makes its way into the ovule. A sperm cell then fertilizes the egg. The zygote that is produced develops into an embryo within the ovule. In time, the embryo matures to a seed. Eventually, the seed falls from the cone and germinates, and the germinating embryo becomes a new pine tree.

The function of dispersal in gymnosperms is assumed by the seeds. Growth of the embryo depends on food supplied from the parent sporophyte, where food-rich tissue surrounds the embryo. The gametophyte generation is little more than a reproductive mechanism in the gymnosperms. Both male and female gametophytes are tiny and entirely dependent on the parent sporophyte.

Vascular plants with protected seeds (angiosperms)

The angiosperms are the most developed and most complex vascular plants. They are the *flowering plants,* of which more than a quarter of a million species have been identified. Almost all vegetables, flowers, fruits, cereals, grains, grasses, and hardwood trees are angiosperms, the dominant life-form on earth today.

Angiosperm means "seed vessel," a reference to the female tissues that enclose the seed. The tissue is **endosperm.** During embryonic development, the endosperm serves as a source of nourishment. In many angiosperms, the endosperm develops into the fruit of the plant. Thus, the protected seed is often found within a fruit. The two most distinguishing features of angiosperms are the flower and the fruit.

The flower of an angiosperm (see Figure 19-1) consists of a ring of modified leaves called **sepals** that enclose and protect the growing flower bud. In some species, the sepals are small and green, while in others they become colored and resemble the petals, the next ring. Flower petals are colorful and are useful in attracting pollinating animals, especially insects. Within the petals are the organs of reproduction: the male stamens and, at the center, the female pistil.

Figure 19-1 The flower of a plant and its structural features.

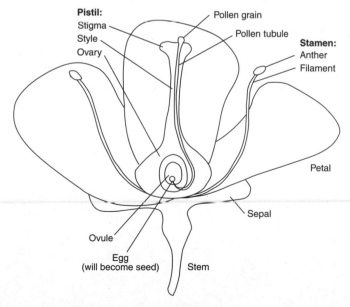

The **stamen** of a flower consists of a thin, stemlike filament and an anther, where haploid pollen grains are produced. Each pollen grain is a male gametophyte containing sperm cells. The **pistil** consists of a sticky *stigma;* a *style,* which is a narrow stalk connecting the stigma to the top of the ovary; and the **ovary,** which is where the ovules are enclosed. In each ovule, a single mother cell divides to produce four cells, one of which will develop into a female gametophyte. The female gametophyte is a sac in which there are eight cells, one of which is the egg cell. The ovary becomes the seed when its egg has been fertilized.

In the angiosperm life cycle, a pollen grain lands on the stigma and produces a pollen tube that grows from the pollen grain down through the style, penetrating the ovary to reach the ovule. One of the sperm cells in the pollen tube fertilizes the egg cell in the ovary to produce a diploid zygote. The zygote becomes the new plant embryo. Another sperm cell fuses with two other cells in the female gametophyte to produce a triploid (three sets of chromosomes) endosperm nucleus, which develops into the endosperm that will feed the growing embryo. The ovule becomes the seed in which the embryo develops, and the ovary ripens into a fruit.

During plant development, the cells become more specialized. In angiosperms, the embryo remains dormant for a while. When the seed germinates, the embryo grows further. Germination may depend on the availability of oxygen, a suitable temperature, or adequate light.

Germination requires the availability of stored food within the seed. In plants called *monocots* (monocotyledon plants), most of the food is stored in the endosperm, in one seed leaf called a *cotyledon* (see Figure 19-2). In *dicots* (dicotyledon plants), most of the food is stored in endosperm in two cotyledons. Among the monocots are grasses, orchids, irises, and lilies. Dicots have two seed leaves (cotyledons) that provide nutrients for the plants as they grow for the seed. Monocots have a single cotyledon. Monocots usually have parallel leaf veins and flower parts occurring in threes. Dicots, in contrast, usually have netlike leaf veins and flower parts in fours or fives. Most of the flowering plants are dicots.

Figure 19-2 Growth of monocotyledon (monocot) and dicotyledon (dicot) plants. Other parts of the young plant are illustrated.

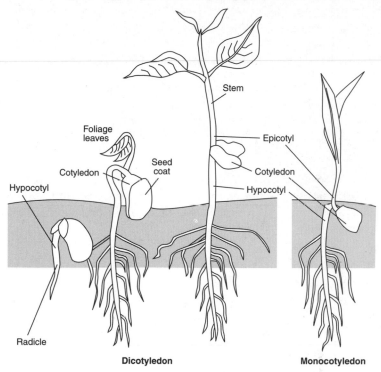

Chapter Checkout

Q&A

1. Each of the following is a group of vascular plants except _____.

   a. gymnosperm
   b. angiosperm
   c. bryophytes

2. Vascular plants such as the conifers (cedars, firs, pines) are members of which of the following groups of vascular plants?

   a. angiosperm
   b. gymnosperm
   c. pteridophyta

3. The pteridophyta produce spores by meiosis, and the spores are stored in cases called _____ on the underside of the leaf, or the _____.

a. ascus, frond
b. sorus, frond
c. sorus, zylem

4. The _____ conducts water and transports minerals upward from the roots whereas the _____ transports sugars and other nutrients from the leaves to the other parts of the plant.

a. phloem, zylem
b. sorus, zylem
c. zylem, phloem

5. In the anatomy of a flower, the stemlike filament and anther is called the _____, a _____ connects the stigma to the ovary, and the ovary is where the _____ are stored.

a. stamen, style, ovules
b. stamen, pistil, zygote
c. pistil, style, ovules

Answers: 1. c **2.** b **3.** b **4.** c **5.** a

Chapter 20

VASCULAR PLANTS: STRUCTURE AND FUNCTION

Chapter Check-In

❑ Studying the vascular system

❑ Differentiating plant tissues

❑ Understanding the functions of roots and stems

The plant world is conveniently separated into two major groups: the nonvascular plants and the vascular plants. The nonvascular plants include the bryophytes, while the vascular plants include the ferns, gymnosperms, and angiosperms (see Chapter 19). The nonvascular plants have no internal transport system. The vascular plants do have such a system, and they are more structurally and functionally complex.

Highly specialized tissues occur in the vascular plants. (A tissue is a group of cells working together to carry out a specialized function.) The tissues are organized into specialized organs called roots, stems, and leaves. The organs all contain the internal vascular system and are connected to one another by the system. The internal vascular system is composed of xylem and phloem.

Different combinations of tissues make up the organs of a vascular plant. In vascular plants, four types of tissues may be found: vascular tissue, ground tissue, dermal tissue, and meristematic tissue.

Vascular Tissue

The vascular tissues include **xylem,** which conducts water and minerals from the roots upward and throughout the plant, and **phloem,** which transports dissolved foods in all directions within the plant.

The main conducting vessels of xylem are the tracheids and the vessels. **Tracheids** are long, thin tubes found in most vascular plants, while **vessels** are large tubes found predominantly in angiosperms. The tracheids and vessels form pipelines that have pores and perforated ends, which allow water and minerals to be conducted from one tube to the next and out to the surrounding tissues. Tracheids and vessels also help support the plant body.

The main conducting cells of phloem are sieve cells and sieve tube members. Both cell types have numerous pores through which substances are exchanged with adjacent cells. Sieve tube members occur in angiosperms, while sieve cells are found in other vascular plants. In angiosperms, small cells called *companion cells* assist the sieve tube members in their functions.

Ground Tissue

The ground tissue of the vascular plant is responsible for storing the carbohydrates produced by the plant. Ground tissue comprises the majority of a young plant and lies between the vascular and dermal tissues.

The major cells of the ground tissue are *parenchyma cells,* which function in photosynthesis and nutrient storage. They have thin walls, many chloroplasts, and large vacuoles, and they form the mass of most leaves, stems, and roots.

Another cell of the ground substance is the *sclerenchyma cell.* Sclerenchyma cells are hollow with strong walls, and they help strengthen the ground tissue.

Dermal Tissue

Dermal tissue functions to protect the plant from injury and water loss. Dermal tissue covers the outside of the plant, except in woody shrubs and trees, which have bark. The most common cell type in dermal tissue is the epidermal cell. Generally, a thin, waxy layer called a *cuticle* covers the epidermal cells and protects them. Other cells in the dermal tissue are guard cells that surround the **stomata,** which are openings in the leaves. Gases and water enter and leave the dermal tissue through the stomata.

Meristematic Tissue

Meristematic tissue is growth tissue and the location of most cell division. It is known as *undifferentiated tissue* because cells in the meristematic tissue will eventually become vascular, ground, or dermal tissue.

Plants generally grow where meristematic tissue is present. At the tips of roots and stems, the meristematic tissue is called the *apical meristem.* The primary growth of the plant occurs in the apical meristem.

Lateral growth in a plant is called secondary growth, which occurs in *lateral meristem* tissue. Woody trees and shrubs display secondary growth when the plants become enlarged and thickened.

Roots

The primary functions of roots in plants are to anchor the plants to the ground and to take in water and minerals from the soil. Substances usually enter the roots by diffusion, but facilitated diffusion may also occur. In addition, roots may be specialized for storage.

Various plants have different types of roots. A *taproot* is a root that grows straight downward and has strong lateral roots growing out of it. Dicots often have taproots. A fibrous root system, consisting of slender, branching roots, occurs in most monocots. Adventitious roots occur on the stems or leaves of some plants, such as corn.

An apical meristem forms the root's tissues. Cells at the tip of the root form the *root cap* that functions mainly for protection. Behind the root cap is the apical meristem and behind the apical meristem is the region of elongation, followed by the region of maturation. Many of the epidermal cells of the root develop extensions called *root hairs* that increase the surface area of the root.

Stems

The stems of vascular plants have several functions, including support of the plant, transport of water and minerals by the vascular system, and generation of energy through photosynthetic cells (in some plants). Some stems also function in food and water storage.

Stems arise in the apical meristem. The outer stem layer is the epidermis, the next layer is composed of vascular tissues, the next is the cortex of parenchyma cells, and at the center of the stem is the **pith.**

In herbaceous plants (for example, clover, potatoes, and wheat) the stem is soft and is composed primarily of meristematic tissue. In woody plants, in contrast, the stems are hard with secondary tissues formed after the primary tissues have been laid down. Secondary tissues arise from the *vascular cambium,* a thin layer of dividing cells between the xylem and phloem.

The vascular cambium is responsible for the lateral growth in the diameter of the woody stem as the cells lay down secondary xylem toward the inside of the stem and secondary phloem toward the outside of the stem. Xylem becomes the wood of the stem, while phloem, together with a tough tissue called **cork,** becomes the **bark** of the stem. Between the phloem and cork is a thin layer of cells—the *cork cambium*—that produces the cork. Annual rings form from xylem tissue.

Monocots and dicots contrast in the construction of stem tissue. In monocot stems, the vascular bundles are scattered throughout the parenchyma. In dicot stems, the vascular bundles are arranged in a ring around the margin of the stem. Most of the interior dicot stem is occupied by pith. In both monocots and dicots, the xylem provides some structural support for the cell.

In woody stems, the apical meristem is embedded in the tip of the stem within a structure called the *terminal bud.* Along the side of the stem are smaller *lateral buds* where new branches and twigs emerge. Leaves generally unfold at intervals along the stem as the terminal bud moves upward. The leaves are attached at points on the stem called *nodes.* Spaces between the nodes are called *internodes.* Openings called *lenticels* are found along woody stems. Lenticels function as pores to permit the exchange of gases between the stem tissue and surrounding air.

Leaves

The leaves are the principal organs of photosynthesis in the vascular plants. The cuticle surrounds the epidermis of the leaf to reduce water loss, while gases pass through pores called stomata. Beneath the upper epidermis of the leaf is a layer of elongated palisade cells. The *palisade cells* contain numerous chloroplasts where photosynthesis takes place. Below the palisade cells is the *spongy mesophyll,* an arrangement of loosely packed cells that also contain chloroplasts for photosynthesis. The air spaces around the cells permit efficient gas exchange to take place during photosynthesis.

Bundles of vascular tissues extend through the leaf and form its veins. The vascular tissue supplies water and nutrients to the photosynthetic cells, and the products of photosynthesis are conducted away from the cells through the phloem. Vascular tissue also runs through the *petiole,* the stalk that connects the leaf to the node of the stem. The broad, flat portion of the leaf is the *blade.*

One of the most important activities in the leaf is the opening and closing of the stomata. These pores regulate the rate of gas exchange, which regulates the rate of photosynthesis. The opening and closing of a stoma is regulated by osmotic pressure within a pair of guard cells. Guard cells are thicker on their inner sides than on the outside, so when the cells are swollen with water, they bow outward, opening the stoma. The pressure exerted on the guard cells to open is called **turgor pressure.** Scientists believe that a low concentration of carbon dioxide and an accumulation of potassium ions in the guard cells instigate their opening. ATP provides the energy for opening and closing the guard cells.

When the stomata are open, the carbon dioxide needed for photosynthesis enters the leaf, while the oxygen gas produced in photosynthesis leaves the leaf. The water produced during photosynthesis also leaves through the stomata. This water loss is called *transpiration.*

Water Movement

The movement of water from the roots to the leaves is a critical function in a plant's life. The flow of water depends upon air pressure, humidity, adhesion, and cohesion. At sea level, normal air pressure can force water up the columns of xylem from the roots to a height of many feet. These columns of water continue to flow upward because water molecules stick to the walls of xylem by *adhesion* and stick to one another by *cohesion.* Water initially moves into the root hair cells by osmosis, because the mineral content of the cells is higher than that of the surrounding environment. Thus, a root pressure is established and extends into the microscopic tubes of the xylem.

At the far end of the system, water either moves out of the plant through the stomata or diffuses into the photosynthetic cells and the spaces around them. This loss of water from the xylem at the far end allows water to flow up the xylem tubes, creating continuous water movement.

In the phloem cells, fluids move in a somewhat similar fashion. In many cases, there is a flow of carbohydrate from the source of production to a depository known as the "sink." For example, sucrose is produced in a photosynthesizing leaf and is transported into phloem tubes. It exits the phloem far away in the root. This is the "source-sink" phenomenon. A constant flow of water is established in the phloem because water follows the carbohydrates by osmosis.

Plant Hormones

The growth and development of many plants are regulated by the activity of plant hormones. **Hormones** are biochemical substances produced in one part of a plant and transported to a different part where they exert a particular effect.

An example of a plant hormone is a series of substances called *auxins.* Auxins increase the length of most plant cells and thereby contribute to the growth and elongation of the plant.

Another plant hormone is *abscisic acid,* which is produced in mature leaves and inhibits growth in developing leaves and germinating seeds. The inhibition occurs during the winter, contributing to the plant's dormancy. Another hormone, *ethylene,* encourages ripening and the dropping of leaves and fruits from the trees. Slight pressure permits the fruit to break loose from the stem.

Two important growth-regulating hormones are the gibberellins and the cytokinins. *Gibberellins* affect plants by stimulating their growth via rapid stem elongation. *Cytokinins* induce plant cells to undergo mitosis; therefore, they encourage increased growth in the roots and stems of plants. They also enhance flowering and stimulate some types of seeds to germinate.

The interactions of hormones and stimuli in the environment often results in a bending or turning response in the plant called a **tropism.** When the plant turns toward a stimulus, the tropism is said to be positive. If the plant turns away from a stimulus, the tropism is negative.

One of the most familiar plant responses is the bending of the stem toward a light source. Light is the stimulus, and the response of the plant is called a **phototropism.** A **geotropism** is a turning of the plant away from or toward the earth. A negative geotropism is a turning away from the earth, such as by a plant stem that grows upward. A positive geotropism is a turning toward the earth, such as in a root that grows downward.

Chapter Checkout

Q&A

1. The main conducting systems of the xylem are the _____ and the _____.

    a. tracheids, sieve cells
    b. vessels, sieve cells
    c. tracheids, vessels

2. The two major cells of the ground tissue are _____ and _____, which function in photosynthesis and nutrient storage.

    a. parenchyma, sclerenchyma
    b. parenchyma, epidermal
    c. epidermal, guard

3. Which of the following is considered to be the principle organ of photosynthesis in the vascular plants?

    a. stems
    b. meristem
    c. leaves

4. The plant hormone _____ inhibits the growth of developing leaves and germinating seeds.

    a. cytokines
    b. abscisic acid
    c. auxin

Answers: 1. c **2.** a **3.** c **4.** b

Chapter 21

ANIMALS: INVERTEBRATES

Chapter Check-In

❏ Recognizing the sponges

❏ Living in marine environments: the jellyfish, mollusks, and sea urchins

❏ Introducing the flatworms, roundworms, and earthworms

❏ Crawling around with the spiders, centipedes, and insects

Animals are multicellular, eukaryotic organisms. They are different from plants in that they take in food and digest it into smaller components. They are heterotrophic rather than autotrophic. In addition to these characteristics, animals are able to move during some point of their life cycle. Also, the primary mode of reproduction in animals is sexual. An animal grows and changes from a single fertilized egg cell into a multicellular organism, passing through various stages of development during its life cycle.

Two major groups of animals exist in the world: the invertebrates and the vertebrates. **Invertebrates** are animals that have no backbones, while **vertebrates** are animals with backbones (see Chapter 22). Among the invertebrates are numerous phyla of animals comprising approximately 95 percent of all the animal species.

Porifera

The phylum Porifera includes a number of simple animals commonly referred to as *sponges*. Sponges filter and consume fine food particles through their pores. Most sponges live attached to rocks, plants, or other animals in marine environments.

The simplest sponges resemble vases or clusters of tubes with irregular shapes. Each sponge has a large, central opening and hundreds of pores in the body wall. Cells lining the cavity remove and digest suspended food particles and water. Waste exits through the central opening.

Sponges are generally without symmetry, and they have a relatively simple cellular organization. The body wall contains a protective layer of flat cells on the outside, an inner layer of flagellated cells, and a gelatinous filling between the two layers. This gelatinous layer contains a group of wandering cells called *amoebocytes*. A sponge is supported by protein that may contain spikes of silicon or calcium compounds. This protein "scaffolding" is typical of the bath sponge.

The cells in the sponge act independently, and there is no evidence of tissue organization. Reproduction occurs by sexual or asexual means. It is possible that sponges evolved from protozoa living in colonies.

Cnidaria

Members of the phylum Cnidaria include hydras, jellyfish, sea corals, and sea anemones. Cnidarians live primarily in marine environments. They have tissue organization and a body plan displaying *radial symmetry;* that is, the organisms are circular with structures that radiate outward. The ends of the structures have tentacles with stinging devices called *cnidoblasts* that help in defense and in trapping food. Digestion of food occurs within the central body cavity called the *gastrovascular cavity.* Cells lining this cavity produce enzymes to break down the food, and the products of digestion are taken into the cells. In addition, the animals have a loose network of nerve cells within their tissue. These nerve cells coordinate the animals' activities.

There are two basic body plans in the cnidarians: a hollow vaselike body called the *polyp,* and an inverted umbrella shape called the *medusa.* Hydras occur as polyps, while jellyfish appear as medusae. A three-layered body wall makes up the outer surface of the polyp or medusa.

Platyhelminthes

Members of the phylum Platyhelminthes are flatworms, such as the planarian. Grubs and tapeworms are other examples of flatworms. Flatworms display *bilateral symmetry;* that is, the left and right halves of the body are mirror images of one another. Another characteristic of the platyhelminthes is cephalization. *Cephalization* means that one end of the animal functions as a head. The "head" contains a mass of nerve cells that acts as a brain and specialized regions for sensing light, chemicals, and pressure.

Flatworms have three distinct layers of tissue, all composed of living cells. They have true organs and organ systems for digestion, movement, excretion, and reproduction. The digestive system consists of a muscular tube with one opening at the mouth. The *excretory system* consists of a network of water-collecting tubules that empty their contents into sacs leading to the exterior.

Movement occurs by the contraction of muscle cells that lie below the epidermis. The contractions are coordinated by signals from a nervous system. Most platyhelminthes have both testes and ovaries, and the organisms pair up to exchange sperm and eggs in fertilization.

Aschelminthes (Nematoda)

The phylum Aschelminthes is also known as Nematoda, and its members are nematodes. The members of this phylum are roundworms, and many are microscopic.

Roundworms have many of the same characteristics of flatworms: for example, bilateral symmetry and cephalization. They also have a digestive tract open at both the mouth and anus. This tract is suspended in a body cavity that is said to be false, so it is called a *pseudocoel.*

Many species of roundworms are free-living organisms that help consume dead plant and animal matter. Several parasitic roundworms cause human diseases, including trichinosis, hookworm disease, and elephantiasis.

Annelida

Members of the phylum Annelida are segmented worms, such as the earthworm. The segmented worms all display bilateral symmetry, cephalization, an open digestive system, segmentation, and a body cavity.

The body cavity is a true body cavity called a *coelom.* This is a fluid-filled space between the innermost cell layer and the outer two cell layers. In this space, the reproductive and digestive organs have evolved into complex structures with complex functions. Also the digestive system is cushioned in the coelom, so the activities of the digestive system take place without interacting with the inner or outer body walls.

Earthworms and other annelids have numerous segments, each separated from the others by internal partitions. Funnel-shaped excretory units called *nephridia,* located in most of the segments, remove water and waste.

Needed water is reabsorbed, and waste material passes out of the body through pores in the skin. Each segment has longitudinal and circular muscles that contract, compressing fluid to form a water-based skeleton called a **hydrostatic skeleton.**

The annelids have a digestive and circulatory system running their entire length. The circulatory system is closed, and blood is circulated by the contraction of several muscular vessels called *hearts.*

The phylum Annelida contains several classes of worms, among them the worms found in mud and sand, the familiar earthworms, and the leeches. All annelids reproduce by a sexual method where sperm and egg cells are released into water for fertilization to take place.

Mollusca

Members of the phylum Mollusca are soft-bodied animals, such as the snail, clam, squid, oyster, and octopus. Generally, these animals live in water. Some secrete a hard shell. Each mollusk has a muscular organ called a *foot* that is used for gripping or creeping over surfaces. The animal has a head with a mouth; a brain or a sense organ; and groups of internal organs for circulation, excretion, respiration, and reproduction. All mollusks have a *mantle,* a thick fold of tissue that covers the internal organs. The space between the mantle and the body organs is the *mantle cavity.*

Unlike the previous animals, the mollusks have a well-developed circulatory system with a chambered heart for pumping blood. The blood passes through gills where oxygen is obtained from water and where carbon dioxide is given off. The circulatory system is said to be "open" because blood passes into open spaces at one point during its circulation.

Most mollusks have well-developed nervous systems with large brains and sense organs. The animals have a closed digestive system with only one opening, and the system is suspended in a true coelom. All animals of the phylum Mollusca have bilateral symmetry.

Three major classes make up the phylum Mollusca. The first class includes the *gastropods,* a group of land snails and slugs. The second class encompasses the *bivalves,* a group of oysters, clams, and other organisms with two shells. The third class is the *cephalopods,* a group of animals with numerous arms and suckers that extend from the "head foot." Octopuses and squids are in this group.

Arthropoda

The largest number of species in the animal kingdom belongs to the phylum Arthropoda. Members of this phylum, called arthropods, include such animals as spiders, ticks, centipedes, lobsters, and insects. All animals in the phylum have an external skeleton, a segmented body, and jointed appendages.

The external body skeleton of arthropods is the **exoskeleton.** It surrounds the animal and provides support. The bodies of arthropods are often divided into distinct regions called the head, thorax, and abdomen. Mouthparts exist in the head region. Antennae, the sense organs, are located on the head also, if they are present. Respiration in the arthropods occurs through microscopic holes in the exoskeleton and body wall. Branching networks called **tracheae** extend from these holes to all parts of an arthropod's body.

Five major classes divide the phylum Arthropoda. The first two classes include the centipedes and millipedes. These animals have multiple appendages and are able to move efficiently on land. Another class of arthropods is the crustaceans. In these animals, the exoskeleton, hardened with calcium salts, serves as a protective shell. Crabs, shrimp, lobster, and crayfish are crustaceans. The arachnids comprise the fourth class of arthropods. Arachnids have a fused head and thorax and usually have four pairs of legs. Within this group are spiders, ticks, mites, and scorpions.

The largest group of arthropods is the insects. Insects live in every conceivable environment on earth and are among the most highly adapted of all animal species. Insects have well-developed organs for various senses, including smell, touch, taste, and hearing. The animals have three pairs of jointed legs, and many have one or two pairs of wings. The body is divided into three parts: head, thorax, and abdomen. The insects include grasshoppers, butterflies, beetles, and cockroaches. In many cases, the embryo develops into an immature form called the *larva.* The larva transforms into a pupa and then becomes an adult. This change is called *metamorphosis.*

Echinodermata

The phylum Echinodermata includes members having an internal skeleton and a water-based pressure system for locomotion. The embryo of echinoderms is different from that of all other invertebrates, and its structure suggests an ancestry to the phylum Chordata.

All echinoderms have spiny skin that helps protect the animals from predators. All are marine creatures, and all the larvae have bilateral symmetry. A ring of nerve tissue surrounds the mouth, and nerves branch off from this tissue to extend throughout the body.

Members of the phylum include sea stars, brittle stars, sea urchins, and sea cucumbers. The animals have the ability to reproduce by regeneration, a process in which a relatively small piece of the animal can regenerate an entire body. All echinoderms have an internal support system called an **endoskeleton** and a large body cavity containing a set of canals called a *water vascular system.*

Chordata

The final phylum of animals is Chordata. All its members, called **chordates,** have bilateral symmetry as well as a head, body cavity, digestive system, and body segmentation.

In addition, chordates have several unique structures. One such structure is the **notochord.** This is a stiff-yet-flexible rod of tissue extending the length of the animal that provides internal support. A second structure is a hollow **nerve chord** (also called a *spinal cord*) that extends the length of the animal just above the notochord. A third characteristic is the presence of *gill slits:* paired openings from the back of the organism's mouth to the outside.

Not all these characteristics exist in the adult form of the chordates. Some exist in the embryonic form. However, all these characteristics exist at some time in the chordate's life cycle. Most biologists agree that chordates evolved from the echinoderms. The special features of the chordates, such as the notochord, spinal cord, and gill slits, are adaptations to the environment.

The chordates are divided into two major groups: the chordate without backbones (the invertebrates) and the chordates with backbones (the vertebrates). The most primitive chordates are **invertebrates.** They include the tunicates and lancelets, both of which lack a backbone but have all the chordate traits. Both of these groups somewhat resemble tadpoles and are found clinging to rocks in marine environments. The vertebrate chordates are discussed in Chapter 22.

Chapter Checkout

Q&A

1. The members of the phylum Mollusca contain a muscular organ called a _____ and a thick fold of tissue that covers the internal organs called a _____.

    a. mantle, foot
    b. foot, coelom
    c. foot, mantle
    d. coelom, mantle

2. The external skeleton of the members of the phylum Arthropoda is called the _____ whereas the sense organs are called the _____.

    a. exoskeleton, tracheae
    b. exoskeleton, antennae
    c. segment, antennae
    d. hydrostatic skeleton, coelom

3. All members of which of the following phylum have bilateral symmetry as well as a head, body cavity, digestive system, and body segmentation?

    a. Echinodermata
    b. Arthropoda
    c. Porifera
    d. Chordata

4. The sponges make up the phylum _____ whereas the jellyfish and sea anemones are members of the phylum _____.

    a. Cnidaria, Porifera
    b. Porifera, Cnidaria
    c. Porifera, Platyhelminthes
    d. Aschelminthes, Porifera

5. All echinoderms contain an internal support system called a(n) _____ and a large body cavity with canals called a(n) _____.

    a. endoskeleton, coelom
    b. endoskeleton, water vascular system
    c. coelom, endoskeleton
    d. endoskeleton, pseudocoel

6. Roundworms are members of the phylum _____ whereas the segmented worms are members of the phylum _____.

   a. Annelida, Nematoda
   b. Nematoda, Annelida
   c. Perifora, Nematoda
   d. Nematoda, Perifora

Answers: 1. c **2.** b **3.** d **4.** b **5.** b **6.** b

Chapter 22

ANIMALS: VERTEBRATES

Chapter Check-In

❑ Characterizing fishes and amphibians

❑ Recognizing reptiles and birds

❑ Defining groups of mammals

All the members of the animal kingdom can be conveniently divided into the invertebrates or vertebrates. Invertebrates are animals that lack a backbone, while vertebrates have a backbone. Invertebrates are found in all the animal phyla (see Chapter 21), including the phylum Chordata.

The phylum Chordata includes animals that have a notochord, a hollow nerve cord, and gill slits. Within this phylum, the vertebrates are classified in the subphylum Vertebrata. There are more than 40,000 living species of vertebrates. Vertebrates are the largest animals ever to inhabit earth; the dinosaurs were vertebrates, as is the blue whale. The vertebrates are divided into several classes, encompassing the fishes, amphibians, reptiles, birds, and mammals.

Fishes

Fishes are aquatic animals with a streamlined shape and a functional tail that allow them to move rapidly through water. Fishes exchange gases with their environment through **gills,** although a few species have lungs that supplement the gills for gas exchange.

One class of fishes called *Agnatha* is made up of species that have no jaws. Lampreys and hagfishes are species within this class. These fishes feed by sucking blood and other body fluids from their prey, usually other fishes.

Another class of fishes is *Placodermi*. The placoderms are now extinct and are recognized as the earliest known fishes with jaws. Placoderms were freshwater bottom-dwellers with a partially bony endoskeleton and bony, armored plates. Another class of fish is *Chondrichthyes*. These fishes have endoskeletons composed entirely of cartilage. The class includes the sharks, skates, and rays. Gas exchanges are made at several prominent, vertical gill openings on either side of the throat.

The bony fishes belong to the class *Osteichthyes*. Most of the familiar fishes living today are bony fishes. They live in the oceans (for example, tuna, mackerel, and herring) and in fresh water (for example, striped bass, trout, and goldfish). The bony fishes prospered during the Devonian Period, which is also called the Age of Fishes. Bony fishes have gills as well as fleshy pectoral and pelvic fins. The ray-finned fishes are the predominant type of bony fishes living today. These fishes have a *swim bladder,* which is a gas-filled sac near the gut that permits a fish to change its buoyancy.

Amphibians

The amphibians are animals that live both on land and in the water. The members of the class *Amphibia* are believed to have evolved from the lobe-finned fishes about 20 million years ago, taking advantage of the higher concentration of oxygen in air than water. They are represented by the frogs, toads, and salamanders.

Amphibians live on land and breathe air to meet their oxygen demands. Amphibians are also able to exchange gases through their lungs, skin, and the inner lining of their mouth. Gas exchange is enhanced by a very efficient circulatory system.

Amphibians remain in moist environments or water to avoid dehydration. Also, amphibians lay their eggs in water because the eggs would quickly dry out on land. Sperm cells are released into the water where they fertilize the egg mass. The early-stage tadpoles lead an aquatic existence and later emerge onto land as adult amphibians.

Reptiles

The reptiles dominated the earth for a period of more than 150 million years. The modern survivors of the Age of Reptiles include the lizards, snakes, crocodiles, alligators, and turtles. In ancient times, the predominant reptiles were the dinosaurs. Reptiles belong to the class *Reptilia*.

Reptiles display a number of adaptations that support their life on land. They have a dry, scaly skin that retards water loss. The structural makeup of their limbs provides better support and allows reptiles to move more quickly than any of the amphibians.

Reproduction in reptiles occurs exclusively on land. The male places sperm into the body of the female, and the embryo develops within an **egg,** which is laid on dry land. Other reptilian characteristics pertain to the respiratory and circulatory systems. The lungs have a greater surface area than that of the amphibians and permit more air to be inhaled. The circulatory system includes a three-chambered heart that separates oxygen-rich and oxygen-poor blood.

Birds

Birds belong to the class *Aves.* These animals have many structural adaptations for flight. For instance, the body is streamlined to minimize air resistance, and the endoskeleton bones are light and hollow. Many of the bones are also fused to provide compact strength. To enable flight, birds also have feathers, which are lightweight adaptations of reptilian scales. Feathers also insulate against the loss of body heat and water.

Birds are **homeothermic,** meaning they are able to maintain a constant body temperature. The rapid pumping of their four-chambered heart and a high blood-flow rate contribute to this characteristic. Insulating feathers also help maintain a constant body temperature.

Mammals

Members of the class *Mammalia* are animals that have hair and nourish their young with milk produced by mammary glands. The presence of body hair or fur helps maintain a constant body temperature in the homeothermic mammals.

Several types of mammals exist: the monotremes, marsupials, and placentals. **Monotremes** are egg-laying mammals that produce milk. The duck-billed platypus and the spiny anteater are examples. They are both found in Australia and probably developed during the geographic isolation of this continent.

Marsupials are mammals whose embryos develop within the mother's uterus for a short period of time before birth. After birth the immature babies crawl into the mother's abdominal pouch where they complete their development. Animals such as the kangaroo, opossum, and koala bear are marsupials.

The **placental mammals** include many familiar animals, such as rabbits, deer, dogs, cats, bats, whales, monkeys, and humans. These mammals have a *placenta:* a nutritive connection between the embryo and the mother's uterine wall. Embryos are attached to the placenta, and they complete their development within the mother's uterus.

Mammals have spread to virtually all environments on earth ranging from the oceans to the deserts. They live underground, on the ground surface, in trees, and in the air. Mammals have a highly developed nervous system, and many have acute senses of smell, hearing, taste, vision, and touch. Mammals rely on memory and learning to guide their activities. They have been able to develop numerous appropriate responses to different environmental situations. They are considered the most successful animals on earth today.

Chapter Checkout

Q&A

1. Fishes exchange gases with the environment through _____.

   a. gills

   b. jaws

   c. lungs

2. Animals that are capable of living on both land and water are members of the _____ class.

   a. Reptile

   b. Mammal

   c. Amphibian

3. The circulatory system of _____ includes a three-chambered heart that separates oxygen-rich and oxygen-poor blood.

   a. amphibians

   b. fishes

   c. reptiles

4. Which of the following is not a type of mammal?

   a. Marsupials

   b. Vertebrata

   c. Monotremes

Answers: 1. a **2.** c **3.** c **4.** b

Chapter 23

NUTRITION AND DIGESTION

Chapter Check-In

❑ Understanding nutritional requirements

❑ Introducing the human digestive system

*N*utrition refers to the activities by which living things obtain raw materials from the environment and transport them into their cells. The cells metabolize these raw materials and synthesize structural components, enzymes, energy-rich compounds, and other biologically important substances. All the elements and compounds taken into a living thing are *nutrients*.

Animals, including humans, are heterotrophic organisms, and their nutrients consist of preformed organic molecules. The organic molecules rarely come in forms that are readily useful, so animals must process the foods into forms that can be absorbed. This processing is called *digestion*.

Nutrition in Animals

The nutritional requirements of most animals are relatively extensive and complex compared with the simple requirements of plants. The nutrients used by animals include carbohydrates, lipids, nucleic acids, proteins, minerals, and vitamins.

■ **Carbohydrates** are the basic source of energy for all animals. Animals obtain their carbohydrates from the external environment (compared with plants, which synthesize carbohydrates by photosynthesis). About one-half to two-thirds of the total calories every animal consumes daily are from carbohydrates. **Glucose** is the carbohydrate most often used as an energy source. This monosaccharide is metabolized during cellular respiration (see Chapter 6), and part of the energy is used to synthesize adenosine triphosphate (ATP). Other useful carbohydrates are maltose, lactose, sucrose, and starch.

■ **Lipids** are used to form cellular and organelle membranes, the sheaths surrounding nerve fibers, and certain hormones. One type of lipid, the fats, are extremely useful energy sources.

■ **Nucleic acids** are used for the construction of deoxyribonucleic acid (DNA), ribonucleic acid (RNA), and ATP. Animals obtain their nucleic acids from plant and animal tissues, especially from cells that contain nuclei. During digestion, the nucleic acids are broken down into nucleotides, which are absorbed into the cells.

■ **Proteins** form the framework of the animal body. Proteins are essential components of the cytoplasm, membranes, and organelles. They are also the major components of muscles, ligaments, and tendons, and they are the essential substances of enzymes. Proteins are composed of 20 kinds of amino acids. Although many amino acids can be synthesized, many others must be supplied in the diet. During digestion, proteins are broken down into their constituent amino acids, which are absorbed into the body.

■ Among the **minerals** required by animals are phosphorus, sulfur, potassium, magnesium, and zinc. Animals usually obtain these minerals when they consume plants. **Vitamins** are organic compounds essential in trace amounts to the health of animals. Vitamins can be water soluble or fat soluble. Water-soluble vitamins must be consumed frequently, while fat-soluble vitamins are stored in the liver in fat droplets. Among the many essential vitamins are vitamin A for good vision, vitamin B for substances used in cellular respiration (FAD, NAD, and coenzyme A), and vitamin D to assist calcium absorption in the body.

Animals obtain their nutrients through a broad variety of feeding patterns. Many animal species, such as sponges, feed on small particles of food that enter their pores. Other aquatic organisms, such as sea cucumbers, wave their tentacles about and trap food on their sticky surfaces. Mollusks, such as clams and oysters, feed by filtering materials through a layer of mucus in their gills. Other animal species, such as certain arthropods, feed exclusively on fluids.

Some animals feed on food masses, and they usually have organs for seizing, chewing, and consuming food. **Herbivores** are animals that eat plants, while **carnivores** are animals that eat other animals. **Omnivores,** which consume both plants and animals, are typified by humans.

Human Digestive System

The human digestive system is a complex process that consists of breaking down large organic masses into smaller particles that the body can use as fuel. The breakdown of the nutrients requires the coordination of several enzymes secreted from specialized cells within the mouth, stomach, intestines, and liver. The major organs or structures that coordinate digestion within the human body include the mouth, esophagus, stomach, small and large intestine, and liver.

Mouth

In the human body, the mouth (oral cavity) is a specialized organ for receiving food and breaking up large organic masses. In the mouth, food is changed mechanically by biting and chewing. Humans have four kinds of teeth: *incisors* are chisel-shaped teeth in the front of the mouth for biting; *canines* are pointed teeth for tearing; and *premolars* and *molars* are flattened, ridged teeth for grinding, pounding, and crushing food.

In the mouth, food is moistened by saliva, a sticky fluid that binds food particles together into a soft mass. Three pairs of **salivary glands**—the parotid glands, the submaxillary glands, and the sublingual glands—secrete saliva into the mouth. The saliva contains an enzyme called *amylase*, which digests starch molecules into smaller molecules of the disaccharide maltose.

During chewing, the tongue moves food about and manipulates it into a mass called a *bolus.* The bolus is pushed back into the pharynx (throat) and is forced through the opening to the esophagus.

Esophagus

The **esophagus** is a thick-walled muscular tube located behind the windpipe that extends through the neck and chest to the stomach. The bolus of food moves through the esophagus by **peristalsis:** a rhythmic series of muscular contractions that propels the bolus along. The contractions are assisted by the pull of gravity.

Stomach

The esophagus joins the stomach at a point just below the diaphragm. A valvelike ring of muscle called the *cardiac sphincter* surrounds the opening to the stomach. The sphincter relaxes as the bolus passes through and then quickly closes.

The *stomach* is an expandable pouch located high in the abdominal cavity. Layers of stomach muscle contract and churn the bolus of food with gastric juices to form a soupy liquid called **chyme.**

The stomach stores food and prepares it for further digestion. In addition, the stomach plays a role in protein digestion. Gastric glands called *chief cells* secrete pepsinogen. Pepsinogen is converted to the enzyme pepsin in the presence of hydrochloric acid. Hydrochloric acid is secreted by *parietal cells* in the stomach lining. The pepsin then digests large proteins into smaller proteins called peptides. To protect the stomach lining from the acid, a third type of cell secretes mucus that lines the stomach cavity. An overabundance of acid due to mucus failure may lead to an ulcer.

Small intestine

The soupy mixture called chyme spurts from the stomach through a sphincter into the small intestine. An adult's small intestine is about 23 feet long and is divided into three sections: the first 10 to 12 inches form the **duodenum;** the next 10 feet form the **jejunum;** and the final 12 feet form the **ileum.** The inner surface of the small intestine contains numerous fingerlike projections called *villi.* Each villus has projections of cells called *microvilli* to increase the surface area.

Most chemical digestion takes place in the duodenum. In this region, enzymes digest nutrients into simpler forms that can be absorbed. Intestinal enzymes are supplemented by enzymes from the **pancreas,** a large, glandular organ lying near the stomach. In addition, bile enters the small intestine from the gall bladder to assist in fat digestion.

The enzymes functioning in carbohydrate digestion include amylase (for starch), maltase (for maltose), sucrase (for sucrose) and lactase (for lactose). For fats, the principal enzyme is lipase. Before this enzyme can act, the large globules of fat must be broken into smaller droplets by bile. *Bile* is a mixture of salts, pigments, and cholesterol that is produced by the liver and stored in the gall bladder, a saclike structure underneath the liver.

Protein digestion is accomplished by several enzymes, including two pancreatic enzymes: trypsin and chymotrypsin. Peptides are broken into smaller peptides, and peptidases reduce the enzymes to amino acids. Nucleases digest nucleic acids into nucleotides in the small intestine also.

Most **absorption** in the small intestine occurs in the jejunum. The products of digestion enter cells of the villi, move across the cells, and enter blood vessels called capillaries. Diffusion accounts for the movement of many nutrients, but active transport is responsible for the movement of glucose and amino acids. The products of fat digestion pass as small droplets of fat into lacteals, which are branches of the lymphatic system.

Absorption is completed in the final part of the small intestine, the ileum. Substances that have not been digested or absorbed then pass into the large intestine.

Large intestine

The small intestine joins the large intestine in the lower right abdomen of the body. The two organs meet at a blind sac called the **cecum** and a small fingerlike process called the **appendix.** Evolutionary biologists believe the cecum and appendix are vestiges of larger organs that may have been functional in human ancestors.

The large intestine is also known as the *colon.* It is divided into ascending, transverse, and descending portions, each about one foot in length. The colon's chief functions are to absorb water and to store, process, and eliminate the residue following digestion and absorption. The intestinal matter remaining after water has been reclaimed is known as *feces.* Feces consist of nondigested food (such as cellulose), billions of mostly harmless bacteria, bile pigments, and other materials. The feces are stored in the rectum and passed out through the anus to complete the digestion process.

Liver

The liver has an important function in processing the products of human digestion. For example, cells of the liver remove excess glucose from the bloodstream and convert the glucose to a polymer called **glycogen** for storage.

The liver also functions in amino acid metabolism. In a process called *deamination,* it converts some amino acids to compounds that can be used in energy metabolism. In doing so, the liver removes the amino groups from amino acids and uses the amino groups to produce urea. Urea is removed from the body in the urine (see Chapter 26). **Fats** are processed into two-carbon units that can enter the Krebs cycle for energy metabolism. The liver also stores vitamins and minerals, forms many blood proteins, synthesizes cholesterol, and produces bile for fat digestion.

Chapter Checkout

Q&A

1. The basic source of energy for all animals is _____.

   a. lipids
   b. carbohydrates
   c. proteins
   d. nucleic acids

2. _____ are animals that eat plants whereas _____ are animals that eat other animals.

   a. herbivores, omnivores
   b. omnivores, herbivores
   c. herbivores, carnivores
   d. omnivores, carnivores

3. The _____ is a thick-walled muscular tube located behind the windpipe that extends through the neck and chest to the stomach.

   a. sphincter
   b. pancreas
   c. bolus
   d. esophagus

4. In the stomach, _____ digests large proteins into smaller peptides.

   a. pepsinogen
   b. hydrochloric acid
   c. pepsin
   d. trypsin

5. Carbohydrates are digested within the _____ whereas the carbohydrate glucose is stored in the _____.

   a. large intestine, liver
   b. small intestine, liver
   c. liver, small intestine
   d. liver, large intestine

Answers: 1. b **2.** c **3.** d **4.** c **5.** b

Chapter 24
GAS EXCHANGE

Chapter Check-In

❑ Directing gas exchange in animals

❑ Respiring through the nose and pharynx

❑ Comparing the trachea and lungs

All living things obtain the energy they need by metabolizing energy-rich compounds, such as carbohydrates and fats. In the majority of organisms, this metabolism takes place by respiration, a process that requires oxygen (see Chapter 6). In the process, carbon dioxide gas is produced and must be removed from the body.

In plant cells, carbon dioxide may appear to be a waste product of respiration too, but because it is used in photosynthesis (see Chapter 5), carbon dioxide may be considered a byproduct. Carbon dioxide must be available to plant cells, and oxygen gas must be removed. Gas exchange is thus an essential process in energy metabolism, and gas exchange is an essential prerequisite to life, because where energy is lacking, life cannot continue.

Mechanisms for Gas Exchange

The basic mechanism of gas exchange is diffusion across a moist membrane. **Diffusion** is the movement of molecules from a region of greater concentration to a region of lesser concentration, in the direction following the concentration gradient. In living systems, the molecules move across cell membranes, which are continuously moistened by fluid.

Simple organisms

Single-celled organisms, such as bacteria and protozoa, are in constant contact with their external environment. Gas exchange occurs by diffusion across their membranes. Even in simple multicellular organisms, such as green algae, their cells may be close to the environment, and gas exchange can occur easily.

In larger organisms, adaptations bring the environment closer to the cells. Liverworts, for instance, have numerous air chambers in the internal environment. The sponge and hydra have water-filled central cavities, and planaria have branches of their gastrovascular cavity that connect with all parts of the body.

Plants

Although plants are complex organisms, they exchange their gases with the environment in a rather straightforward way. In aquatic plants, water passes among the tissues and provides the medium for gas exchange. In terrestrial plants, air enters the tissues, and the gases diffuse into the moisture bathing the internal cells.

In the leaf of the plant, an abundant supply of carbon dioxide must be present, and oxygen from photosynthesis must be removed. Gases do not pass through the cuticle of the leaf; they pass through pores called **stomata** in the cuticle and epidermis. Stomata are abundant on the lower surface of the leaf, and they normally open during the day when the rate of photosynthesis is highest. Physiological changes in the surrounding guard cells account for the opening and closing of the stomata (see Chapter 20).

Animals

In animals, gas exchange follows the same general pattern as in plants. Oxygen and carbon dioxide move by diffusion across moist membranes. In simple animals, the exchange occurs directly with the environment. But with complex animals, such a mammals, the exchange occurs between the environment and the blood. The blood then carries oxygen to deeply embedded cells and transports carbon dioxide out to where it can be removed from the body.

Earthworms exchange oxygen and carbon dioxide directly through their skin. The oxygen diffuses into tiny blood vessels in the skin surface, where it combines with the red pigment **hemoglobin.** Hemoglobin binds loosely to oxygen and carries it through the animal's bloodstream. Carbon dioxide is transported back to the skin by the hemoglobin.

Terrestrial arthropods have a series of openings called **spiracles** at the body surface. Spiracles open into tiny air tubes called **tracheae,** which expand into fine branches that extend into all parts of the arthropod body.

Fishes use outward extensions of their body surface called gills for gas exchange. **Gills** are flaps of tissue richly supplied with blood vessels. As a fish swims, it draws water into its mouth and across the gills. Oxygen diffuses out of the water into the blood vessels of the gill, while carbon dioxide leaves the blood vessels and enters the water passing by the gills.

Terrestrial vertebrates such as amphibians, reptiles, birds, and mammals have well-developed respiratory systems with lungs. Frogs swallow air into their lungs, where oxygen diffuses into the blood to join with hemoglobin in the red blood cells. Amphibians can also exchange gases through their skin. Reptiles have folded lungs to provide increased surface area for gas exchange. Rib muscles assist lung expansion and protect the lungs from injury.

Birds have large air spaces called *air sacs* in their lungs. When a bird inhales, its rib cage spreads apart and a partial vacuum is created in the lungs. Air rushes into the lungs and then into the air sacs where most of the gas exchange occurs. This system is the birds' adaptation to the rigors of flight and their extensive metabolic demands.

The lungs of mammals are divided into millions of microscopic air sacs called **alveoli.** Each alveolus is surrounded by a rich network of blood vessels for transporting gases. In addition, mammals have a dome-shaped diaphragm that separates the thorax from the abdomen, providing a separate chest cavity for breathing and pumping blood. During inhalation, the diaphragm contracts and flattens to create a partial vacuum in the lungs. The lungs fill with air, and gas exchange follows.

Human Respiratory System

The human respiratory system consists of a complex set of organs and tissues that capture oxygen from the environment and transport the oxygen into the lungs. The organs and tissues that comprise the human respiratory system include the nose and pharynx, the trachea, and the lungs.

Nose and pharynx

The respiratory system of humans begins with the nose, where air is conditioned by warming and moistening. Bone partitions separate the nasal cavity into chambers, where air swirls about in currents. Hairs and hairlike cilia trap dust particles and purify the air.

The nasal chambers open into a cavity at the rear of the mouth called the **pharynx** (throat). From the pharynx, two tubes called *Eustachian tubes* open to the middle ear to equalize air pressure there. The pharynx also contains tonsils and *adenoids,* which are pockets of lymphatic tissue used to trap and filter microorganisms.

Trachea

After passing through the pharynx, air passes into the windpipe, or **trachea.** The trachea has a framework of smooth muscle with about 16 to 20 open rings of cartilage shaped like a C. These rings give rigidity to the trachea and ensure that it remains open.

The opening to the trachea is a slitlike structure called the **glottis.** A thin flap of tissue called the **epiglottis** folds over the opening during swallowing and prevents food from entering the trachea. At the upper end of the trachea, several folds of cartilage form the **larynx,** or voicebox. In the larynx, flaplike pairs of tissues called *vocal cords* vibrate when a person exhales and produce sounds.

At its lower end, the trachea branches into two large **bronchi** (singular, *bronchus*). These tubes also have smooth muscle and cartilage rings. The bronchi branch into smaller **bronchioles,** forming a bronchial "tree." The bronchioles terminate in the air sacs known as **alveoli.**

Lungs

Human lungs are composed of approximately 300 million alveoli, which are cup-shaped sacs surrounded by a capillary network. Red blood cells pass through the capillaries in single file, and oxygen from each alveolus enters the red blood cells and binds to the hemoglobin. In addition, carbon dioxide contained in the plasma and red blood cells leaves the capillaries and enters the alveoli when a breath is taken. Most carbon dioxide reaches the alveoli as bicarbonate ions, and about 25 percent of it is bound loosely to hemoglobin.

When a person inhales, the rib muscles and diaphragm contract, thereby increasing the volume of the chest cavity. This increase leads to reduced air pressure in the chest cavity, and air rushes into the alveoli, forcing them to expand and fill. The lungs passively obtain air from the environment by this process. During exhalation, the rib muscles and diaphragm relax, the chest cavity area diminishes, and the internal air pressure increases. The compressed air forces the alveoli to close, and air flows out.

The nerve activity that controls breathing arises from impulses transported by nerve fibers passing into the chest cavity and terminating at the rib muscles and diaphragm. These impulses are regulated by the amount of carbon dioxide in the blood: A high carbon-dioxide concentration leads to an increased number of nerve impulses and a higher breathing rate.

Chapter Checkout

Q&A

1. Gases pass through pores called _____ in plants.

   a. cuticle
   b. spiracles
   c. gills
   d. stomata

2. Fishes exchange gases using _____ whereas the arthropods used air tubes called _____.

   a. gills, tracheae
   b. tracheae, lungs
   c. gills, lungs
   d. tracheae, alveoli

3. Mammalian lungs are divided into millions of air sacs called _____.

   a. tracheae
   b. alveoli
   c. bronchioles
   d. diaphragm

4. Several folds of cartilage form the voice box, which is called the _____.

   a. bronchi
   b. alveoli
   c. larynx
   d. pharynx

Answers: 1. d **2.** a **3.** b **4.** c

Chapter 25

BLOOD AND CIRCULATION

Chapter Check-In

❑ Circulating through plants and animals

❑ Introducing the components of circulation

❑ Defending against foreign antigens

Single-celled organisms are in constant contact with their environments, obtaining nutrients and oxygen directly across the cell surface. The same holds true for small and simple plants and animals, such as algae, bryophytes, sponges, cnidarians, and flatworms. Larger and more complex plants and animals require methods for transporting materials to and from cells far removed from the external environment. These organisms have evolved transport systems.

Circulatory Systems

Circulatory systems function to transport materials to and from cells throughout an organism. Organisms ranging from plants to animals have different nutritional requirements. Due to these differences, various species have evolved distinct circulatory processes to assist with their specific transport needs.

Plants

Transport systems are found in the vascular plants. Vascular networks provide intercellular communication in terrestrial plants. The systems consist of tubelike connective tissues organized into xylem and phloem. Xylem transports water and minerals in the plants, while phloem transports food materials and hormones (see Chapter 20).

Xylem and phloem tissues are grouped in arrangements called **vascular bundles.** In monocot plants, the vascular bundles are scattered throughout the parenchyma tissue in no particular pattern. In dicot plants, the vascular bundles occur in a circle around a central region of pith. In woody dicot plants, new xylem forms on the inside of the cambium each season; the old xylem forms the annual rings of the plant.

Animals

In animals, the transport system is generally called a **circulatory system** because the transport fluid flows through a circuit. Most animals have one or more organs called *hearts* for pumping the fluid. The channels through which the fluid flows are the **arteries** (which lead away from the heart), the **veins** (which lead to the heart), and the **capillaries** (the microscopic blood vessels between arteries and veins).

In animals, such as earthworms, the circulatory system consists of blood, channels, and five pulsating vessels that function as hearts, rushing blood through the vessels to all the earthworm's body parts. Gases bind to hemoglobin in the blood.

Terrestrial arthropods have an open circulatory system. A tubelike heart pumps blood into a dorsal blood vessel, which empties into the arthropod's body cavity, or *hemocoel.* Contractions of the body muscles gradually move blood back toward the animal's heart.

All vertebrates have a single, strong muscular heart to pump the blood. In fishes, the blood accumulates in a thin-walled receiving chamber called the **atrium.** The blood then passes through a valve into a pumping chamber, a **ventricle.** The ventricle contracts and forces blood out to the gills, where gas exchange occurs, and from there to the remainder of the body. Veins bring the blood back to the atrium.

Amphibians, such as frogs, have a three-chambered heart. The heart has a right and a left atrium and a single ventricle. In reptiles, a muscular septum exists between the two sides of the ventricle, creating a primitive four-chambered heart. Birds have a more sophisticated four-chambered heart than reptiles. One ventricle pumps blood to the lungs for gas exchange, while the second pumps oxygen-rich blood to the rest of the body tissues. Mammals also have a four-chambered heart.

Human Circulatory System

The human circulatory system functions to transport blood and oxygen from the lungs to the various tissues of the body. The heart pumps the blood throughout the body. The lymphatic system is an extension of the human circulatory system that includes cell-mediated and antibody-mediated immune systems. The components of the human circulatory system include the heart, blood, red and white blood cells, platelets, and the lymphatic system.

Heart

The human heart is about the size of a clenched fist. It contains four chambers: two atria and two ventricles. Oxygen-poor blood enters the **right atrium** through a major vein called the **vena cava.** The blood passes through the **tricuspid valve** into the **right ventricle.** Next, the blood is pumped through the **pulmonary artery** to the lungs for gas exchange. Oxygen-rich blood returns to the **left atrium** via the **pulmonary vein.** The oxygen-rich blood flows through the **bicuspid (mitral) valve** into the **left ventricle,** from which it is pumped through a major artery, the **aorta.** Two valves called **semilunar valves** are found in the pulmonary artery and aorta.

The ventricles contract about 70 times per minute, which represents a person's pulse rate. Blood pressure, in contrast, is the pressure exerted against the walls of the arteries. Blood pressure is measured by noting the height to which a column of mercury can be pushed by the blood pressing against the arterial walls. A normal blood pressure is a height of 120 millimeters of mercury during heart contraction (*systole*), and a height of 80 millimeters of mercury during heart relaxation (*diastole*). Normal blood pressure is usually expressed as "120 over 80."

Coronary arteries supply the heart muscle with blood. The heart is controlled by nerves that originate on the right side in the upper region of the atrium at the sinoatrial node. This node is called the *pacemaker.* It generates nerve impulses that spread to the atrioventricular node where the impulses are amplified and spread to other regions of the heart by nerves called **Purkinje fibers.**

Blood

Blood is the medium of transport in the body. The fluid portion of the blood, the **plasma,** is a straw-colored liquid composed primarily of water. All the important nutrients, the hormones, and the clotting proteins as well as the waste products are transported in the plasma. Red blood cells and white blood cells are also suspended in the plasma. Plasma from which the clotting proteins have been removed is **serum.**

Red blood cells

Red blood cells are **erythrocytes.** These are disk-shaped cells produced in the bone marrow. Red blood cells have no nucleus, and their cytoplasm is filled with hemoglobin.

Hemoglobin is a red-pigmented protein that binds loosely to oxygen atoms and carbon dioxide molecules. It is the mechanism of transport of these substances. (Much carbon dioxide is also transported as bicarbonate ions.) Hemoglobin also binds to carbon monoxide. Unfortunately, this binding is irreversible, so it often leads to carbon-monoxide poisoning.

A red blood cell circulates for about 120 days and is then destroyed in the spleen, an organ located near the stomach and composed primarily of lymph node tissue. When the red blood cell is destroyed, its iron component is preserved for reuse in the liver. The remainder of the hemoglobin converts to bilirubin. This amber substance is the chief pigment in human bile, which is produced in the liver.

Red blood cells commonly have immune-stimulating polysaccharides called **antigens** on the surface of their cells. Individuals having the A antigen have blood type A (as well as anti-B antibodies); individuals having the B antigen have blood type B (as well as anti-A antibodies); individuals having the A and B antigens have blood type AB (but no anti-A or anti-B antibodies); and individuals having no antigens have blood type O (as well as anti-A and anti-B antibodies).

White blood cells

White blood cells are referred to as **leukocytes.** They are generally larger than red blood cells and have clearly defined nuclei. They are also produced in the bone marrow and have various functions in the body. Certain white blood cells called **lymphocytes** are essential components of the immune system (discussed later in this chapter). Other cells called **neutrophils** and **monocytes** function primarily as **phagocytes;** that is, they attack and engulf invading microorganisms. About 30 percent of the white blood cells are lymphocytes, about 60 percent are neutrophils, and about 8 percent are monocytes. The remaining white blood cells are **eosinophils** and **basophils.** Their functions are uncertain; however, basophils are believed to function in allergic responses.

Platelets

Platelets are small disk-shaped blood fragments produced in the bone marrow. They lack nuclei and are much smaller than erythrocytes. Also known technically as **thrombocytes,** they serve as the starting material for **blood clotting.** The platelets adhere to damaged blood vessel walls, and thromboplastin is liberated from the injured tissue. Thromboplastin, in turn, activates other clotting factors in the blood. Along with calcium ions and other factors, thromboplastin converts the blood protein prothrombin into thrombin. Thrombin then catalyzes the conversion of its blood protein fibrinogen into a protein called *fibrin,* which forms a patchwork mesh at the injury site. As blood cells are trapped in the mesh, a blood clot forms.

Lymphatic system

The lymphatic system is an extension of the circulatory system consisting of a fluid known as lymph, capillaries called lymphatic vessels, and structures called lymph nodes. **Lymph** is a watery fluid derived from plasma that has seeped out of the blood system capillaries and mingled with the cells. Rather than returning to the heart through the blood veins, this lymph enters a series of one-way **lymphatic vessels** that return the fluid to the circulatory system. Along the way, the ducts pass through hundreds of tiny, capsulelike bodies called **lymph nodes.** Located in the neck, armpits, and groin, the lymph nodes contain cells that filter the lymph and phagocytize foreign particles.

The **spleen** is composed primarily of lymph node tissue. Lying close to the stomach, the spleen is also the site where red blood cells are destroyed. The spleen serves as a reserve blood supply for the body.

The lymph nodes are also the primary sites of the white blood cells called lymphocytes. The body has two kinds of lymphocytes: **B-lymphocytes** and **T-lymphocytes.** Both of these cells can be stimulated by microorganisms or other foreign materials called **antigens** in the blood. Antigens are picked up by phagocytes and lymph and delivered to the lymph nodes. Here, the lymphocytes are stimulated through a process called the **immune response.**

Certain antigens, primarily those of fungi and protozoa, stimulate the T-lymphocytes. After stimulation, these lymphocytes leave the lymph nodes, enter the circulation, and proceed to the site where the antigens of microorganisms were detected. The T-lymphocytes interact with the microorganisms cell to cell and destroy them. This process is called **cell-mediated immunity.**

B-lymphocytes are stimulated primarily by bacteria, viruses, and dissolved materials. On stimulation, the B-lymphocytes revert to large antibody-producing cells called **plasma cells.** The plasma cells synthesize proteins called **antibodies,** which are released into the circulation. The antibodies flow to the antigen site and destroy the microorganisms by chemically reacting with them in a highly specific manner. The reaction encourages phagocytosis, neutralizes many microbial toxins, eliminates the ability of microorganisms to move, and causes them to bind together in large masses. This process is called **antibody-mediated immunity.** After the microorganisms have been removed, the antibodies remain in the bloodstream and provide lifelong protection to the body. Thus, the body becomes immune to specific disease microorganisms.

Chapter Checkout

Q&A

1. Oxygen-poor blood enters the right atrium of the human heart through the _____ and is pumped through the _____ to the lungs.

   a. vena cava, pulmonary vein
   b. pulmonary vein, vena cava
   c. vena cava, pulmonary artery
   d. pulmonary artery, vena cava

2. The fluid portion of the blood is called _____.

   a. serum
   b. plasma
   c. semilunar
   d. basophil

3. True or False: Red blood cells do not contain a nucleus.

4. Which of the following are not types of white blood cells?

   a. neutrophils
   b. eosinophils
   c. lymphocytes
   d. thrombocytes

5. Red blood cells are destroyed in the _____, and the _____ are the primary sites of white blood cells.

   a. lymph nodes, spleen
   b. spleen, lymph nodes
   c. plasma, lymph nodes
   d. spleen, platelets

Answers: 1. c **2.** b **3.** True. **4.** d **5.** b

Chapter 26

EXCRETION AND HOMEOSTASIS

Chapter Check-In

❑ Maintaining steady-state equilibrium

❑ Detailing the human excretory system

❑ Controlling kidney function

The excretion process is concerned with the removal of waste products from the animal body. Dependent upon the process of excretion is the relatively stable internal environment of the organism, a concept known as **homeostasis.** Homeostasis means "staying the same." It refers to a relatively stable internal environment, or a steady-state equilibrium that exists internally in a healthy organism, despite changes in the external environment. The excretory system plays a major role in homeostasis.

Excretory Systems

Because one-celled organisms are in constant contact with their environment, they do not need excretory organs. However, multicellular organisms need a mechanism to carry waste products from cells to the external environment. Flatworms, such as planaria, have a series of excretory cells, called *flame cells*. Flame cells contain cilia that direct water and metabolic wastes to enter the cells and to pass into excretory canals. The excretory canals join with other canals to form excretory tubules. Fluid from the excretory tubules leaves the body through pores.

In earthworms, members of the phylum Annelida, the excretory system consists of structural units called *nephridia*. Each nephridium contains a ciliated tunnel that leads to a long, coiled tubule, which leads to a bladderlike sac (a primitive bladder). Fluid moves from the internal environment into the funnel. As fluid passes through the tubule, cells in the tubular lining absorb useful compounds such as glucose, amino acids, and salts. The remaining materials constitute metabolic waste, and they are passed into the bladderlike sac. The sac later opens through a pore in the earthworm's skin where the waste products are discharged.

Insects have a series of tubules for excretion called *Malpighian tubules*. Fluid enters at the upper end of the tubules and passes down their entire length. The cells in the tubular walls reabsorb precise amounts of water, salts, and other materials to maintain delicate balance within the insect tissues. The tubules eventually lead to an insect's intestine where waste products are removed.

Human Excretory System

The human excretory system functions to remove waste from the human body. This system consists of specialized structures and capillary networks that assist in the excretory process. The human excretory system includes the kidney and its functional unit, the nephron. The excretory activity of the kidney is modulated by specialized hormones that regulate the amount of absorption within the nephron.

Kidneys

The human kidneys are the major organs of bodily excretion (see Figure 26-1). They are bean-shaped organs located on either side of the backbone at about the level of the stomach and liver. Blood enters the kidneys through **renal arteries** and leaves through **renal veins.** Tubes called **ureters** carry waste products from the kidneys to the **urinary bladder** for storage or for release.

The product of the kidneys is **urine,** a watery solution of waste products, salts, organic compounds, and two important nitrogen compounds: uric acid and urea. **Uric acid** results from nucleic acid decomposition, and **urea** results from amino acid breakdown in the liver. Both of these nitrogen products can be poisonous to the body and must be removed in the urine.

Figure 26-1 Details of the human excretory system. Position and allied structures of the kidneys (top). A cross section of the kidney showing the two major portions (left). Details of the nephron, the functional unit of the kidney (right).

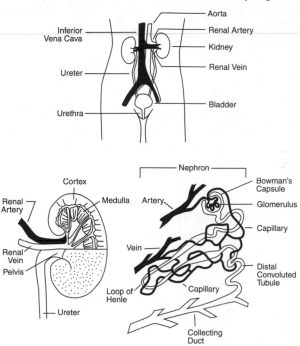

Nephron

The functional and structural unit of the kidney is the **nephron.** The nephron produces urine and is the primary unit of homeostasis in the body. It is essentially a long tubule with a series of associated blood vessels. The upper end of the tubule is an enlarged cuplike structure called the **Bowman's capsule.** Below the Bowman's capsule, the tubule coils to form the *proximal tubule,* and then it follows a hairpin turn called the *loop of Henle.* After the loop of Henle, the tubule coils once more as the *distal tubule.* It then enters a *collecting duct,* which also receives urine from other distal tubules.

Within the Bowman's capsule is a coiled ball of capillaries known as a **glomerulus.** Blood from the renal artery enters the glomerulus. The force of the blood pressure induces plasma to pass through the walls of the glomerulus, pass through the walls of the Bowman's capsule, and flow into the proximal tubule. Red blood cells and large proteins remain in the blood.

After plasma enters the proximal tubule, it passes through the coils, where usable materials and water are reclaimed. Salts, glucose, amino acids, and other useful compounds flow back through tubular cells into the blood by active transport. Osmosis and the activity of hormones assist the movement. The blood fluid then flows through the loop of Henle into the distal tubule. Once more, salts, water, and other useful materials flow back into the bloodstream. Homeostasis is achieved by this process: A selected amount of hydrogen, ammonium, sodium, chloride, and other ions maintain the delicate salt balance in the body.

The fluid moving from the distal tubules into the collecting duct contains materials not needed by the body. This fluid is referred to as *urine.* Urea, uric acid, salts, and other metabolic waste products are the main components of urine. The urine flows through the ureters toward the urinary bladder. When the bladder is full, the urine flows through the **urethra** to the exterior.

Control of kidney function

The activity of the nephron in the kidney is controlled by a person's choices and environment as well as hormones. For example, if a person consumes large amounts of protein, much urea will be in the blood from the digestion of the protein. Also, on a hot day, a body will retain water for sweating and cooling, so the amount of urine is reduced.

Humans produce a hormone called *antidiuretic hormone (ADH),* also known as *vasopressin,* which is secreted by the posterior lobe of the pituitary gland. It regulates the amount of urine by controlling the rate of water absorption in the nephron tubules.

Some individuals suffer from a condition in which they secrete very low levels of ADH. The result is excessive urination and a disease called *diabetes insipidus.* Another unrelated form of diabetes, *diabetes mellitus,* is more widespread. Persons with this disease produce insufficient levels of insulin. Insulin normally transports glucose molecules into the cells. But when insulin is not available, the glucose remains in the bloodstream. The glucose is removed from the bloodstream in the nephron; to dilute the glucose, the nephron removes large amounts of water from the blood. Thus, the urine tends to be plentiful.

Hormones from the cortex of the adrenal glands also control the content of urine. These hormones promote reabsorption of sodium and chloride ions in the tubules. Thus, they affect the water balance in the body, because water flows in the direction of high sodium and chloride content.

Chapter Checkout

Q&A

1. Flatworms contain a series of excretory cells called _____.

    a. nephridia cells

    b. flame cells

    c. Malpighian cells

    d. Nephral tubes

2. Blood enters the human kidney through the _____ and leaves through the _____.

    a. renal arteries, renal veins

    b. renal veins, renal arteries

    c. renal arteries, ureter

    d. ureter, renal veins

3. Capillaries within the Bowman's capsule make up the _____.

    a. nephron

    b. ureter

    c. distal tube

    d. glomerulus

4. Antidiuretic hormone regulates the rate of which of the following processes?

    a. water excretion

    b. water absorption

    c. insulin absorption

    d. insulin excretion

Answers: 1. b **2.** a **3.** d **4.** b

Chapter 27

SUPPORT AND MOVEMENT

Chapter Check-In

❑   Comparing skeletal systems

❑   Studying movement in animal cells

❑   Contracting your muscles

Movement is one of the essential features of living things. Cellular movement is observed in one-celled amoebas, ciliates, and flagellates. Flagella whip about to produce cellular motion, while cilia beat synchronously to propel a cell.

In animals, movement is essential for locating food and seeking mates. The movement process is centered in the muscle cell, which contracts and relaxes. The contraction yields great force, which is applied against a surface by means of a skeleton.

Skeletons in Animals

Skeletal systems provide structure or protection to a variety of organisms. A water-based skeleton provides the structure necessary for movement in worms. The hard external skeleton provides a protective mechanism for many organisms but also assists in movement of insects. The internal skeleton present in many animals provides the structural network for support, protection, and movement.

Hydrostatic skeleton

Many animals have a water-based skeleton, or **hydrostatic skeleton.** Hydrostatic skeletons do not contain hard structures, such as bone, for muscles to pull against. Rather, the muscles surround a fluid-filled body cavity. In a worm, for example, movement occurs when muscle cells contract, and the contractions squeeze internal fluid (the hydrostatic skeleton) against the skin, causing the worm to stiffen and the body to shorten and widen. The squirming motion of a worm also depends on a hydrostatic skeleton.

Exoskeleton

A second type of skeleton, the **exoskeleton**, exists in arthropods and mollusks. In mollusks, the exoskeleton is a hard, protective, outer covering; an example is a clamshell. When a clam's muscles contract, they close the shell rapidly, creating a spurt of water that propels the clam. In arthropods, the exoskeleton also provides protection and movement. Usually, wings are attached by muscles in the hard body surface, which provides a foundation for the muscle contractions. Muscle contractions raise and lower the wings, allowing flight.

Endoskeleton

Most vertebrates have an internal skeleton called the **endoskeleton,** which is a framework of bones and cartilage (see Figure 27-1) that serves as a point of attachment for muscle. The endoskeleton thus transmits the force of muscle contractions. The endoskeleton also provides support for the body (for example, the legs) and protection (the skull).

Bone tissue contains concentric rings of tissue in which bone cells called *osteoblasts* produce the inorganic materials (fibers and matrix) of bone. Much of this material is calcium phosphate, formed from calcium and phosphorus delivered by the blood. Living, mature bone cells called *osteocytes* are also located in the bone. Bone-destroying cells called *osteoclasts* break down bone, thus providing a turnover of bone material needed in other areas. The combination of bone cells and bone tissue comprises a unit called a *Haversian system.* Blood vessels and nerves also exist within the Haversian system.

Bones come together to form a joint, which may be immovable, such as in the **sutures** of the skull, or movable, such as in the joints of the elbow and shoulder. In a movable joint, a capsule of synovial fluid provides lubrication. In a joint, tough, fibrous tissues, known as **ligaments,** link bones to one another. Connective tissues, called **tendons,** attach muscles to bones.

Figure 27-1 The human skeleton showing the major bones of the body.

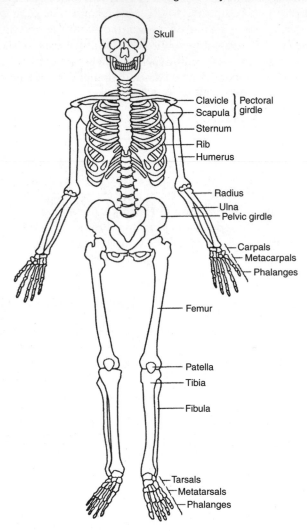

Skull

Clavicle ⎤ Pectoral
Scapula ⎦ girdle

Sternum

Rib

Humerus

Radius

Ulna
Pelvic girdle

Carpals
Metacarpals
Phalanges

Femur

Patella

Tibia

Fibula

Tarsals
Metatarsals
Phalanges

Human Muscle

Muscle is composed of thousands of muscle fibers, each composed of a single muscle cell. As shown in Figure 27-2, a muscle cell contains a series of ultramicroscopic filaments called **myofibrils.** Each myofibril is a muscle cell that contains units called **sarcomeres.** Sarcomeres contain thick microfilaments composed of the protein **myosin.** Sarcomeres also contain thin microfilaments composed of the protein **actin.** The actin and myosin filaments are arranged parallel to one another, with the myosin filaments' molecular "heads" protruding toward the actin filaments. In skeletal muscle, the overlapping actin and myosin filaments give the muscle fiber a banded, or striated, appearance. Hence the muscle is **striated muscle.**

Figure 27-2 Anatomical structure of the muscle.

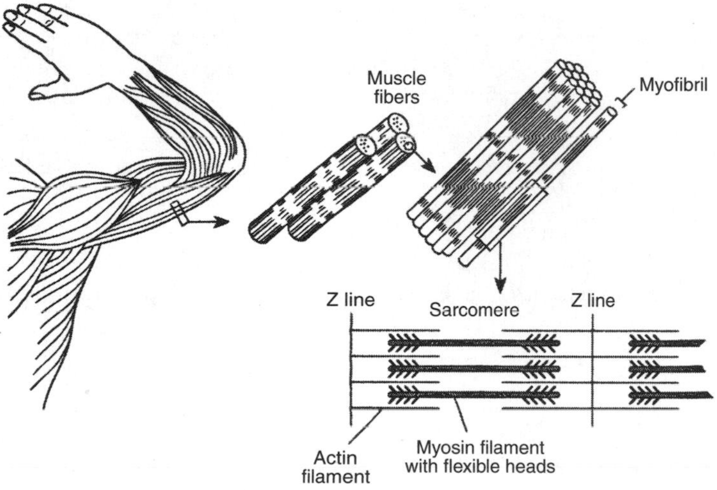

Muscle contraction

When a nerve impulse arrives at the muscle cells, it passes across the neuromuscular junction and enters the muscle cell membrane, which is known as the **sarcolemma.** The impulse spreads across the muscle cell and enters its cytoplasm, which is called *sarcoplasm.* The nerve impulse causes the actin filaments to slide across the surface of the myosin filaments. The sliding filaments pull together the ends of the muscle cell, thereby causing it to contract. The sliding filaments require that calcium ions and energy in the form of ATP be available. Two proteins called *tropomyosin* and *troponin* also function in the contraction. Cross-bridges hold the filaments together as the muscle contracts.

After the contraction has taken place, the energy to sustain the contraction is used up, and the cross-bridges break. The filaments then slide back to their original position, and the muscle cell relaxes. There is no partial contraction of the muscle cell. Contraction is an all-or-none phenomenon.

Energy for contraction

Adenosine triphosphate (ATP) supplies the energy for muscle contraction. The reactions of glycolysis, the Krebs cycle, and the electron transport system normally produce ATP during cell respiration (see Chapter 6). During normal activity, ATP is regenerated as it is used up during muscle contraction. When a person engages in strenuous activity, however, ATP is quickly used up and creatin phosphate is used for energy. Creatin phosphate transfers its energy to new ATP molecules, which then function as additional energy sources.

When creatin phosphate is used up, muscle cells obtain their energy solely from the process of glycolysis. Because little or no oxygen is available, the metabolism is anaerobic. Under these conditions, two molecules of ATP are obtained per molecule of glucose metabolized. The pyruvic acid that forms is converted to lactic acid in the muscle. Lactic acid prevents over-exertion of the muscle because as lactic acid accumulates, the person experiences fatigue. The fatigue induces the person to stop exerting the muscles and breathe deeply. This breathing provides plentiful supplies of oxygen to satisfy the oxygen debt. Lactic acid is converted back to pyruvic acid, which is then metabolized through the Krebs cycle and electron transport system to provide a new supply of ATP and creatin phosphate.

Types of muscle

The human body has three major types of muscle. The muscle type discussed previously is *striated muscle* because the fibers' overlapping actin and myosin filaments give it a banded appearance (see Figure 27-2). This muscle is found in the limbs and is also called **skeletal muscle.** It operates under voluntary control and so is additionally known as **voluntary muscle.**

The second muscle type is **smooth muscle,** which has few actin and myosin filaments; therefore, it has few striations. Smooth muscle is found in the linings of the blood vessels, along the gastrointestinal tract, in the respiratory tract, and in the urinary bladder. Because it operates without voluntary control, it is sometimes called **involuntary muscle.**

The third muscle type is **cardiac muscle,** which is found in the heart. It has striations because it has multiple actin and myosin filaments, but it is involuntary muscle. The actin and myosin filaments in cardiac muscle exist as intertwined branches that form a conducting network for nerve impulses.

Chapter Checkout

Q&A

1. Mature bone cells are called _____ and bone-destroying cells are called _____.

    a. osteocytes, osteoclasts
    b. osteoclasts, osteocytes
    c. osteoblasts, osteoclasts
    d. osteoclasts, osteoblasts

2. In joints, bones are linked to each other by fibrous tissues called _____.

    a. tendons
    b. muscles
    c. sutures
    d. ligaments

3. Energy for muscle contraction is supplied by _____.

    a. creatin phosphate
    b. lactic acid
    c. ATP
    d. tropomyosin

4. Skeletal muscle is _____ whereas smooth muscle is _____.

    a. involuntary, voluntary
    b. voluntary, striated
    c. involuntary, striated
    d. voluntary, involuntary

Answers: 1. a **2.** d **3.** c **4.** d

Chapter 28

CHEMICAL COORDINATION

Chapter Check-In

❑ Coordinating hormone secretion

❑ Mapping the human endocrine system

The animal body has two levels of coordination: nervous coordination (see Chapter 29) and chemical coordination. Chemical coordination is centered in a system of glands known as **endocrine glands.** These glands are situated throughout the animal body and include such organs as the pancreas, thyroid gland, and adrenal gland. The glands secrete **hormones,** a series of chemical substances composed of protein or sterol lipids.

Hormones bring about changes that help coordinate body systems in a general way. For example, the pancreas secretes insulin, which facilitates the passage of glucose into all body cells for use in energy metabolism. Another example is thyroxine, a thyroid gland secretion that regulates overall body metabolism. In contrast to chemical coordination, the nervous system coordinates functions in the animal body on a more localized level as it delivers nerve impulses to contract body muscles or regulate gland activities.

The endocrine glands secrete their hormones into the bloodstream, where the blood carries the hormones to the target organs. Because the endocrine glands have no ducts, they are often called **ductless glands.** Other glands of the body (such as the enzyme-secreting salivary glands) deliver their enzymes via ducts and are referred to as **exocrine glands.**

The structure and physiology of hormones and endocrine glands is relatively similar in all animals; the emphasis in this chapter is on the human endocrine system.

Human Endocrine System

The human endocrine system modulates several processes of the body by the function of hormones. The endocrine system secretes hormones that control how bodily functions work. Thus, the human endocrine system watches over and coordinates all the systems of the body by the use of hormones.

Pituitary gland

The pituitary gland is located at the base of the human brain. The gland consists of two parts: the anterior lobe (adenohypophysis) and the posterior lobe (neurohypophysis).

The anterior lobe secretes at least seven hormones. One hormone, the human growth hormone (HGH), promotes body growth by accelerating protein synthesis. This hormone is also known as *somatotropin*. A deficiency of the hormone results in dwarfism; an oversecretion results in gigantism.

Another hormone of the anterior pituitary is prolactin, also called *lactogenic hormone (LH)*. This hormone promotes breast development and milk secretion in females. A third hormone is thyroid-stimulating hormone (TSH). The function of TSH is to control secretions of hormones from the thyroid gland. A fourth hormone is adrenocorticotropic hormone (ACTH). This hormone controls the secretion of hormones from the adrenal glands.

There are three more hormones produced in the anterior lobe of the pituitary gland. The first is follicle-stimulating hormone (FSH). In females, FSH stimulates the development of a follicle, which contains the egg cell; in males, the hormone stimulates sperm production. The next hormone is luteinizing hormone (LH). In females, LH completes the maturation of the follicle and stimulates the formation of the corpus luteum, which temporarily secretes female hormones. In males, LH is interstitial cell-stimulating hormone (ICSH), which stimulates the production of male hormones in the testes. The final hormone is melanocyte-stimulating hormone (MSH), which stimulates production of the pigment melanin.

The posterior pituitary gland stores and then releases two hormones that are produced in the hypothalamus of the brain. The first hormone is antidiuretic hormone (ADH). This hormone stimulates water reabsorption in the kidneys. It is also called *vasopressin.* The second hormone is oxytocin, which stimulates contractions in the muscles of the uterus during birth.

Thyroid gland

The **thyroid gland** lies against the pharynx at the base of the neck. It consists of two lateral lobes connected by an isthmus. The gland produces thyroxine, a hormone that regulates the rate of metabolism in the body. It also produces a second hormone, calcitonin, which regulates the level of calcium in the blood.

Thyroxine production depends on the availability of iodine. A deficiency of iodine causes thyroid gland enlargement, a condition called *goiter.* An undersecretion of thyroxine results in a condition known as *cretinism* (dwarfism with abnormal body proportions and possible mental retardation). In adults, an undersecretion results in *myxedema* (physical and mental sluggishness). Thyroxine oversecretion results in a high metabolic rate and Graves' disease.

Parathyroid glands

The **parathyroid glands** are located on the posterior surfaces of the thyroid gland. They are tiny masses of glandular tissue that produce parathyroid hormone, also called *parathormone.* Parathyroid hormone regulates calcium metabolism in the body by increasing calcium reabsorption in the kidneys, and by increasing the uptake of calcium from the digestive system.

Adrenal glands

The **adrenal glands** are two pyramid-shaped glands lying atop the kidneys. The adrenal glands consist of an outer portion, the cortex, and an inner portion, the **medulla.**

The adrenal cortex secretes a family of steroids called **corticosteroids.** The two main types of steroid hormones are mineralocorticoids and glucocorticoids. Mineralocorticoids, such as aldosterone, control mineral metabolism in the body. They accelerate mineral reabsorption in the kidney. Mineralocorticoid secretion is regulated by ACTH from the pituitary gland. Glucocorticoids, such as cortisol and cortisone, control glucose metabolism and protein synthesis in the body. Glucocorticoids are also anti-inflammatory agents.

The adrenal medulla produces two hormones: **epinephrine** (adrenaline) and **norepinephrine** (noradrenaline). Epinephrine increases heart rate, blood pressure, and the blood supply to skeletal muscle. Epinephrine functions in stressful situations to promote the fight–flight response. Norepinephrine intensifies the effects of epinephrine. Both hormones prolong and intensify the effects of the sympathetic nervous system.

Pancreas

The pancreas is located just behind the stomach. Its endocrine portion consists of cell clusters called the **islets of Langerhans.**

The pancreas produces two hormones: insulin and glucagon. Insulin is a protein that promotes the passage of glucose molecules into the body cells and regulates glucose metabolism. In the absence of insulin, glucose is removed from the blood and excreted in the kidney, a condition called *diabetes mellitus.* Diabetes mellitus is characterized by glucose in the urine, heavy urination, excessive thirst, and a generally sluggish body metabolism.

The second pancreatic hormone, **glucagon,** stimulates the breakdown of glycogen to glucose in the liver. It also releases fat from the adipose tissue so the fat can be used for the production of carbohydrates.

Other endocrine glands

Among the other endocrine glands are the ovaries and testes. The **ovaries** secrete **estrogens,** which encourage the development of secondary female characteristics. The **testes** secrete **androgens**, which promote secondary male characteristics. Testosterone is an important androgen.

The **pineal gland** is a tiny gland in the midbrain. Its functions are largely unknown, but it seems to regulate mating behaviors and day–night cycles. The **thymus gland** is located in the neck tissues. It secretes **thymosins,** which influence the development of the T-lymphocytes of the immune system.

Prostaglandins are hormones secreted by various tissue cells. These hormones produce their effects on smooth muscles, on various glands, and in reproductive physiology. **Erythropoietin** is a hormone produced by the kidney cells. Erythropoietin functions in the production of red blood cells. **Gastrin** and **secretin** are hormones produced by digestive glands to influence digestive processes.

Chapter Checkout

Q&A

1. The anterior pituitary gland produces hormones such as _____ whereas the posterior pituitary gland produces hormones such as _____.

   a. prolactin, calcitonin
   b. vasopressin, calcitonin
   c. prolactin, vasopressin
   d. vasopressin, prolactin

2. The hormone produced in the adrenal medulla that increases the heart rate, blood pressure, and blood supply to skeletal muscles is called _____.

   a. norepinephrine
   b. epinephrine
   c. calcitonin
   d. vasopressin

3. _____ promotes the passage of glucose into cells, whereas _____ induces the breakdown of glycogen.

   a. thyroxine, insulin
   b. oxytocin, insulin
   c. insulin, oxytocin
   d. insulin, glucagon

4. _____ promotes breast development and milk secretion in females.

   a. luteinizing hormone
   b. lactogenic hormone
   c. epinephrine
   d. estrogen

Answers: 1. c **2.** b **3.** d **4.** b

Chapter 29

NERVOUS COORDINATION

Chapter Check-In

❑ Coordinating the human nervous system

❑ Sensing the external environment

Nervous coordination enables an organism's rapid response to an external or internal stimulus. Characteristic of animals only, nervous coordination is the function of the nervous system. The receptors for nervous coordination are generally located in the sense organs at the body surface, while the response in nervous coordination generally involves a gland or muscle. The function of coordination is accomplished by means of a set of signals conducted along a series of nerve cells.

Animal Nerve Cells

Animal nerve cells are specialized cells called neurons. Depending upon function, these cells can be divided into sensory neurons, interneurons, and motor neurons. These three types of nerve cells coordinate with each other to receive external stimuli and to transmit the impulse to muscles or glands of the body for an appropriate response to the stimulus.

Neuron

The **neuron** is the nerve cell. (Approximately 12 billion neurons exist in the human body, the great majority of them in the brain and spinal cord.) The main portion of the neuron is the cell body. Protruding from the **cell body** are one or more short extensions called **dendrites** and one long extension called the **axon.** Axons are covered by a fatty layer of material called the **myelin sheath.** Bundles of axons bound together are referred to as a **nerve.**

There are three types of neurons in animals: sensory neurons, interneurons, and motor neurons. **Sensory neurons** receive stimuli from the external environment; **interneurons** (or association neurons) connect sensory and motor neurons and carry stimuli in the brain and spinal cord; **motor neurons** transmit impulses from the brain and spinal cord to the muscle or gland that will respond to the stimulus. The neurons are supported, protected, and nourished by cells of the nervous system known as **glial cells.** Together with extracellular tissue, the glial cells make up the **neuroglia.**

Nerve impulse

The nerve impulse is an electrochemical event that occurs within the neuron. In an inactive neuron, the cytoplasm is negatively charged with respect to the outside of the cell. This difference in electrical charge is maintained by the active transport of sodium ions out of the cytoplasm. A cell in this state is said to have a **resting potential,** and it is polarized.

A nerve impulse is generated when the difference in electrical charge disappears. This occurs when a stimulus contacts the tip of a dendrite and increases the permeability of the cell membrane to sodium ions. The ions rush back into the cytoplasm, and the difference in electrical charges disappears. This creates a pulse of electrochemical activity called the **nerve impulse.** A neuron displaying a nerve impulse is said to have an **action potential.** The cell is depolarized.

More specifically, the influx of sodium ions into the neuron cytoplasm activates the adjacent portion of the cell membrane to admit sodium ions also. Successively, the adjacent areas of the neuron lose their differences of electrical charge, and a wave of depolarization is generated in the neuron. This wave of depolarization is the nerve impulse. After the wave of depolarization has passed, the neuron reestablishes the difference in charges by pumping potassium ions out of the cytoplasm, and then pumping sodium ions in.

Synapse

The nerve impulse passes down the dendrite, through the cell body, and down the axon. At the end of the axon, the impulse encounters a fluid-filled space separating the end of the axon from the dendrite of the next neuron or from a muscle cell. This space is the **synapse.** A synapse located at the junction of a neuron and muscle fiber is a neuromuscular junction.

As the impulse reaches the end of the axon, it induces changes in the cell membrane and the release of chemical substances called **neurotransmitters,** for example, acetylcholine. Molecules of neurotransmitter accumulate in the synapse and increase the membrane permeability of the next dendrite. This causes an influx of sodium ions, and a new nerve impulse is generated. After the nerve impulse has swept down the next dendrite, the neurotransmitters in the synapse are destroyed.

Reflex arc

The **reflex arc,** the simplest unit of nervous activity, involves the detection of a stimulus in the environment by sensory nerve endings, followed by impulses that travel via the sensory neurons to the spinal cord. Here the impulses synapse with interneurons, and the interneurons generate impulses to respond to the stimulus. The impulses travel along the motor neurons to muscles or glands that respond appropriately.

In some cases, a reflex arc involves an interpretation. For this activity, interneurons transmit impulses up the spinal cord to the conscious area of the brain, where an analysis occurs.

Human Central Nervous System

The human nervous system may be conveniently subdivided into two divisions: the central nervous system (the brain and spinal cord) and the peripheral nervous system (the nerves extending to and from the central nervous system).

Spinal cord

The **spinal cord** of the central nervous system is a white cord of tissue passing through the bony tunnel made by the vertebrae. The spinal cord extends from the base of the brain to the bottom of the backbone. Three membranes called *meninges* surround the spinal cord and protect it. The outer tissue of the spinal cord is white (white matter), while the inner tissue is gray (gray matter).

Thirty-one pairs of projections called **nerve roots** extend out along each side of the spinal cord. The nerve roots are sites of axons belonging to sensory and motor neurons. A central canal in the spinal cord carries cerebrospinal fluid, which provides for the nutrition and gaseous needs of the cord tissue. The neurons of the spinal cord serve as a coordinating center for the reflex arc and a connecting system between the peripheral nervous system and the brain.

Brain

The brain of the central nervous system is the organizing and processing center. It is the site of consciousness, sensation, memory, and intelligence. The brain receives impulses from the spinal cord and from 12 pairs of cranial nerves coming from and extending to the senses and to other organs. In addition, the brain initiates activities without environmental stimuli.

Two major hemispheres, the left and the right hemispheres, make up the tissue of the brain. The outer portion of the brain consists of gray matter, while the inner portion is white matter. Three major portions of the brain are recognized: the hindbrain, the midbrain, and the forebrain.

The **hindbrain** consists of the medulla, cerebellum, and pons. The **medulla** is the swelling at the tip of the brain that serves as the passageway for nerves extending to and from the brain. The cerebellum lies adjacent to the medulla and serves as a coordinating center for motor activity; that is, it coordinates muscle contractions. The **pons** is the swelling between the medulla and midbrain. The pons acts as a bridge between various portions of the brain.

The **midbrain** lies between the hindbrain and forebrain. It consists of a collection of crossing nerve tracts and is the site of the reticular formation, a group of fibers that arouse the forebrain when something unusual happens.

The **forebrain** consists of the cerebrum, the thalamus, the hypothalamus, and the limbic system. The **cerebrum** contains creases and furrows called convolutions that permit the cerebral hemisphere to accommodate more than 10 billion cells. Each hemisphere of the cerebrum has four lobes, and activities such as speech, vision, movement, hearing, and smell occur in these lobes. Higher mental activities such as learning, memory, logic, creativity, and emotion also occur in the cerebrum.

The **thalamus** serves as an integration point for sensory impulses, while the **hypothalamus** synthesizes hormones for storage in the pituitary gland. The hypothalamus also appears to be a control center for such visceral functions as hunger, thirst, body temperature, and blood pressure. The **limbic system** is a collection of structures that ring the edge of the brain and apparently function as centers of emotion.

Human Peripheral Nervous System

The **peripheral nervous system** is a collection of nerves that connect the brain and spinal cord to other parts of the body and the external environment. It is subdivided into the sensory somatic system and the autonomic nervous system.

The **sensory somatic system** carries impulses from the external environment and the senses. It consists of 12 pairs of cranial nerves and 31 pairs of spinal nerves. The sensory somatic system permits humans to be aware of the outside environment and react to it voluntarily.

The **autonomic nervous system** works on an involuntary basis. It consists of two groups of motor neurons and a set of knotlike groups of cell bodies called **ganglia.** Motor neurons extend to and from the ganglia to the body organs. One subdivision of the autonomic nervous system is the **sympathetic nervous system.** Impulses propagated in this system prepare the body for an emergency. They cause the heartbeat to increase, the arteries to constrict, the pupils to dilate, and other changes to take place. The other subdivision is the **parasympathetic nervous system.** Impulses in this system return the body to normal after an emergency has occurred.

Human Senses

The senses are organs that connect the nervous system to the external environment. They are the sources of stimuli that cause a response in the nervous system, and they are the sources of all information to the human body.

Eye

In the human eye, the nerve cells are located in a single layer called the **retina,** located along the back wall of the eye. Light rays enter the eye through a curved, transparent structure called the cornea, then pass through the pupil, an opening in the eyeball. The iris regulates the size of the pupil. Next, the lens focuses the light on the retina, which contains two types of light-sensitive cells, **rod cells** and **cone cells,** which detect light. Cone cells, which detect color, are concentrated in the central portion of the retina, while rod cells, which permit vision in dim light, are concentrated at the edge of the retina. A light-sensitive pigment called **rhodopsin** functions in the detection of light.

From the eye, a series of impulses is generated for transmission to the brain. The **optic nerve** carries these impulses. The region of keenest vision, the *fovea,* is located at the center of the retina. When vision is poor, light is not focusing on the fovea, and corrective lenses are prescribed.

Ear

The ear is the organ of hearing in humans. The outer ear funnels vibrations to the **eardrum,** or tympanic membrane, which transmits the vibrations to three inner earbones: the *malleus* (hammer), the *incus* (anvil), and the *stapes* (stirrup). These bones transmit the vibrations to the inner ear where the receptor of hearing, the cochlea, is located.

The **cochlea** is a snail-like series of coiled tubes within the skull. As the earbones vibrate, they push and pull a membrane at one end of the cochlea, causing fluid within the tubules to vibrate. The vibrations are detected by sensitive hair cells, and nerve impulses are generated. The **auditory nerve** carries the impulses to the brain for interpretation.

Taste and smell

Specialized receptor cells called chemoreceptors transmit taste and smell. **Chemoreceptors** of the human tongue distinguish four different tastes: sweet, sour, salty, and bitter. In the human nose, chemoreceptors detect a variety of scents, including minty, floral, musky, putrid, and pungent.

In both taste and smell, chemoreceptors are stimulated by molecules and ions that reach the tongue and nose. Liquid materials affect the chemoreceptors in the taste buds of the tongue, while gaseous molecules affect the chemoreceptors in the upper reaches of the nose. The **olfactory nerve** carries nerve impulses from the nose to the brain for interpretation.

Other senses

The other senses of the body include receptors for touch, pain, temperature, and balance. Touch and pain receptors, called **Pacinian corpuscles,** are located in the skin, muscles, and tendons. The sense of balance is centered in the semicircular canals of the inner ear. Visceral senses include stretch receptors in the muscles as well as carbon dioxide receptors in the arteries.

Chapter Checkout

Q&A

1. _____ neurons receive stimuli from the external environment whereas _____ neurons transmit impulses from the brain to muscles or glands.

   a. motor, sensory
   b. sensory, interneurons
   c. sensory, motor

2. _____ provides nutrition and gaseous needs of the cord tissue.

   a. cerebrospinal fluid
   b. nerve roots
   c. reflex arc

3. The _____ consists of the cerebrum, thalamus, hypothalamus, and the limbic system.

   a. hindbrain
   b. forebrain
   c. midbrain

4. The _____ nervous system permits humans to voluntarily react to the outside environment whereas the _____ nervous system functions on an involuntary basis.

   a. parasympathetic, sensory somatic
   b. peripheral, sympathetic
   c. sensory somatic, autonomic

Answers: 1. c **2.** a **3.** b **4.** c

Chapter 30

REPRODUCTION

Chapter Check-In

❑ Understanding the male and female reproductive systems

❑ Fertilizing an egg

❑ Developing a fetus

Reproduction is an essential process for the survival of a species. The functions of the reproductive systems are to produce reproductive cells, the **gametes,** and to prepare the gametes for fertilization. In addition, the male reproductive system delivers the gametes to the female reproductive tract. The female reproductive organs nourish the fertilized egg cell and provide an environment for its development into an embryo, a fetus, and a baby.

Human Reproduction

Human reproduction takes place by the coordination of the male and female reproductive systems. In humans, both males and females have evolved specialized organs and tissues that produce haploid cells, the sperm and the egg. These cells fuse to form a zygote that eventually develops into a growing fetus. A hormonal network is secreted that controls both the male and female reproductive systems and assists in the growth and development of the fetus and the birthing process.

Male reproductive system

The male reproductive organs are the testes (or testicles). The testes are two egg-shaped organs located in a pouch called the **scrotum** outside the body. In the scrotum, the temperature is a few degrees cooler than body temperature. The testes develop in the abdominal cavity before birth, and then descend to the scrotum.

Sperm production in the testes takes place within coiled passageways called **seminiferous tubules.** Within the walls of these tubules, primitive cells called **spermatogonia** undergo a series of changes and proceed through meiosis to yield **sperm cells.** Each human sperm cell has 23 chromosomes.

The sperm cells mature in a tube called the **epididymis.** The epididymis is located along the surface of the testes. The hormone that stimulates sperm cell production is the follicle-stimulating hormone (FSH). A second hormone important in reproduction is the interstitial cell-stimulating hormone (ICSH), which acts on interstitial cells located between the seminiferous tubules. The interstitial cells secrete male hormones, including testosterone. The male hormones regulate the development of secondary male characteristics.

The organ responsible for carrying the sperm cells to the female is the penis. Within the penis, the sperm cells are carried in a tube, the **urethra.** During periods of sexual arousal, the penis becomes erect as blood fills its spongelike tissues. The sperm cells are mixed with secretions from the prostate gland, seminal vesicles, and Cowper's glands. These secretions and the sperm cells constitute the **semen.**

Female reproductive system

The organs of female reproduction include the **ovaries,** two oval organs lying within the pelvic cavity, and adjacent to them, two **Fallopian tubes.** Also known as *oviducts,* the Fallopian tubes are the passageways that egg cells enter after release from the ovaries. The Fallopian tubes lead to the **uterus** (womb), a muscular organ in the pelvic cavity. The inner lining, called the *endometrium,* thickens with blood and tissue in anticipation of a fertilized egg cell. If fertilization fails to occur, the endometrium degenerates and is shed in the process of menstruation.

The opening at the lower end of the uterus is a constricted area called the **cervix.** The tube leading from the cervix to the exterior is a muscular organ called the **vagina.** During periods of sexual arousal, the vagina receives the penis and the semen. The sperm cells in the semen pass through the cervix and uterus into the Fallopian tubes, where fertilization takes place.

In the human female, egg cell production begins before birth, when about 2 million primitive cells known as **oogonia** accumulate in the ovaries. These oogonia are formed in the early stages of meiosis. After the age of puberty, the oogonia develop into primary **oocytes** and then into **egg** cells at a rate of one per month. Egg cell production occurs by the process of meiosis.

The egg cells develop within the ovary in a cluster of cells called the **Graafian follicle,** which secretes female hormones called **estrogens** that regulate the development of secondary female characteristics. Egg cell development within the follicle requires approximately 14 days. The development is controlled by two hormones: follicle-stimulating hormone (FSH) and luteinizing hormone (LH). Both hormones are secreted by the anterior lobe of the pituitary gland.

During the 14 days of egg cell development, the endometrium increases its supply of blood and nutrients in anticipation of a fertilized egg cell. On about the 14th day, the release of the egg cell from the follicle takes place. This process is called **ovulation.** The egg cell is swept into the Fallopian tube and begins to move toward the uterus. Meanwhile, the follicle is changed into a mass of cells known as the **corpus luteum.** The LH stimulates the conversion. The corpus luteum then secretes the hormone **progesterone,** which together with estrogens continues to regulate the buildup of tissue in the endometrium and inhibit contractions of the uterus.

The egg cell remains alive in the Fallopian tubes for 24 to 72 hours. If fertilization by a sperm cell fails to occur, the egg cell moves toward the uterus, and the corpus luteum begins to degenerate. This degeneration causes the level of progesterone and estrogen to drop off. Within two weeks, the hormone level declines to a point where it cannot inhibit contractions of the uterus. Uterine contractions then occur, and the endometrium is released in the process of **menstruation.** Follicle development begins again, but in the opposite ovary.

Fertilization and development

For fertilization to occur, sperm cells must be released in the vagina during the period that the egg cell is alive. The sperm cells move through the uterus into the Fallopian tube, where one sperm cell may fertilize the egg cell. The fertilization brings together 23 chromosomes from the male and 23 chromosomes from the female, resulting in the formation of a **fertilized egg cell** with 46 chromosomes. The fertilized cell is a **zygote.**

The zygote undergoes mitosis to form two identical cells that remain attached. This takes place about 36 hours after fertilization. Mitosis then occurs more frequently. Soon a solid ball of cells, a **morula,** results. Morula formation occurs about six days after fertilization. During that time the cells are moving through the Fallopian tube. Within the next two days, a hollow ball of cells called a **blastocyst** forms. The blastocyst enters the uterus. At one end of the blastocyst, a group of cells called the **inner cell mass** continues to develop.

About eight days after fertilization, the blastocyst implants itself in the endometrium of the uterus. During implantation, the outer cells take root in the endometrium. This outer layer of cells, called the **trophoblast,** gives rise to projections that form vessels. These vessels merge with the maternal blood vessels to form the **placenta.** The trophoblast also develops into three membranes: the amnion, the chorion, and the yolk sac membrane.

The inner cell mass undergoes changes to form three germ layers known as the ectoderm, the mesoderm, and the endoderm. The **ectoderm** becomes the skin and nervous system, the **mesoderm** becomes the muscles and other internal organs, and the **endoderm** becomes the gastrointestinal tract. The embryo is formed at about the fourth week when all the organs of the body have taken shape.

Embryonic stage

At the age of four weeks, the embryo is about the size of a pea. A primitive heart is beating, the head is defined with rudimentary eyes and ears, and tiny bumps represent arms and legs. The embryo also contains a primitive nervous system, and the head has begun to enlarge. A cartilage skeleton has appeared, and muscles have taken shape.

By the end of eight weeks, the embryo is somewhat human looking. Facial features are evident, and most of the organs are well developed. From this point onward, development consists chiefly of growth and maturation. The embryo is about 1.5 inches in length. Henceforth it is known as a **fetus.**

Nourishment of the embryo, and then the fetus, is accomplished through the **placenta.** The maternal and embryonic blood supplies meet at this organ, but the blood does not mix. Instead, diffusion accounts for the passage of gases, nutrients, and waste products across the membranous barriers. The placenta is also an endocrine gland because it secretes estrogen and progesterone to continue to inhibit follicle development and maintain the integrity of the endometrium. As the embryo becomes a fetus, it moves away from the placenta, and a length of tissue called the **umbilical cord** becomes its source of attachment to the maternal blood supply.

Fetal development

During the third month, the fetus definitely resembles a human, but the head is relatively large. During the ensuing months, the remainder of the body increases in size proportionally. Cartilage is replaced by bone, and the reproductive organs develop.

During the fourth month, the length of the fetus increases to about 6 inches. The heartbeat can be heard through the mother's abdominal wall, and the fetus moves about. Distinctive movements can be felt at the fifth month, and by the sixth month the fetus weighs almost 2 pounds. By the end of six months, the fetus might be able to survive outside the mother's body, but it would have little fat in its skin so temperature control would be a problem. By the end of the ninth month, the fetus has an average length of about 20 inches and a typical weight of 6 to 8 pounds.

Birth

The birth process, which is complex, is regulated by various hormones. When diminished levels of progesterone remove the inhibition on uterine contractions, the uterine muscles contract. The posterior pituitary gland releases the hormone **oxytocin,** which stimulates further contractions.

During the first stage of labor, the cervix opens and the baby descends into the birth canal, or vagina. By this time, the amniotic sac has broken. In the second stage of labor, the baby passes through the birth canal, assisted by painful uterine contractions. The baby's head normally appears first, but in a breech birth the buttocks may appear first. In the third stage of birth, the placenta (afterbirth) is delivered.

Chapter Checkout

Q&A

1. Sperm production takes place in the _____.

   a. scrotum
   b. seminiferous tubules
   c. prostate

2. Egg cells develop in a cluster of cells called the _____ that secretes the hormone _____.

   a. cervix, luteinizing hormone
   b. vagina, estrogen
   c. Graafian follicle, estrogen

3. The placenta is considered a(n) _____ gland because it secretes estrogen and progesterone.

   a. adrenal

   b. endocrine

   c. parathyroid

4. During development, the _____ becomes the skin whereas the _____ becomes the gastrointestinal tract.

   a. mesoderm, ectoderm

   b. ectoderm, endoderm

   c. endoderm, ectoderm

Answers: 1. b **2.** c **3.** b **4.** b

Chapter 31
ECOLOGY

Chapter Check-In

❑ Occupying a defined area

❑ Defining a community

❑ Interacting to form ecosystems

Ecology is the discipline of biology that is concerned primarily with the interaction between organisms and their environments. There are many levels of organization in the world of living things, and each level has features not displayed by any other. The important levels are the population, community, ecosystem, and biosphere.

Populations

A **population** is a group of individuals belonging to one species usually occupying a defined area. Populations of living things interact with other populations of their own kind, with populations of other species, and with physical aspects of their environment.

A population's growth proceeds until reaching certain environmental limits. When a population has reached the maximum size that the environment can support, the environment is said to have reached its **carrying capacity.** The growth of a population cannot exceed the carrying capacity of its environment for long.

The growth of a population passes through stages. There is an initial lag period of minimal growth, followed by an exponential growth period of maximum population growth. The curbs that limit population growth may include the crowding of a population, an example of a density-dependent population. Density-dependent curbs include disease, competition, predation, and territoriality. Another curb, occurring in a density-independent population, takes place when there is an external limiting factor. Density-independent curbs result from climatic fluctuations, such as temperature changes.

Communities

Communities of plants, animals, and other organisms may be found in such places as a desert, a salt marsh, or a forest. Within a **community,** each population of organisms has a habitat and niche. The habitat is the physical place where the organisms live, while the niche is the role that the population plays in the life of the community. The niche is the population's function and position in the ecosystem, and it reflects the population's relationship to other populations. A population's niche is defined by how and when it reproduces, what time of the day and year it is most active, what climatic factors it can withstand, what it eats, and so forth. The competition exclusion principle suggests that two species cannot occupy the same niche in the same place at the same time.

Two populations living together in a community in a close and permanent association is **symbiosis.** If the symbiotic relationship is mutually beneficial, it is known as **mutualism.** Lichens represent an example of mutualism. If one population receives a benefit from an association, while the other is neither benefited nor harmed, the symbiosis is **commensalism.** Humans and the bacteria of their intestines exist commensally. Another type of symbiosis is **parasitism,** in which one population benefits while the other is harmed. The microorganisms that cause human disease are considered parasites. A final type of symbiosis is **synergism.** In this instance, two populations accomplish together what neither population could accomplish on its own.

Populations within a community may interact; indeed, one population may capture and feed on the other. Such a relationship is **predation.** Predators have more than one prey species, but they normally feed upon the most abundant prey species available. As a prey population decreases, the predator switches to a more abundant species. This change causes fluctuations in population sizes.

Natural selection favors the most efficient predator, while also favoring the prey that can escape predation. Among the adaptations that help prey to escape are poisonous toxins, chemical adaptations, warning coloration (camouflage), and mimicry.

In a community, the orderly and predictable replacement of populations over a given period of time in a given area is *succession,* which is primary succession when new populations are established in new habitats. In secondary succession, communities are established on a site previously occupied by a population. In each succession, certain populations dominate and then decline, to be superseded by new dominant populations. A community at the last stage of succession is the climax community.

Ecosystems

Interactions between communities and their physical environments form systems known as **ecosystems.** One of the major phenomena underlying an ecosystem is the flow of energy. Because photosynthesizing organisms trap the energy in an ecosystem, they are **producers.** Because certain organisms in the community meet their energy needs by feeding on these producers, they are **consumers.** Producers are usually autotrophs, while consumers are usually heterotrophs. Primary consumers feed directly on plants, while secondary consumers (carnivores) feed on the animals that eat the plants. The energy flow forms a one-way pattern; the energy is used for metabolic reactions and the remainder is given off as heat.

The transfer of food energy from producers to consumers is the **food chain.** Many food chains intertwine in a complex manner to form a **food web. Decomposers,** the organisms of decay, are found at each level of the food chains and food webs. The organisms of decay are usually bacteria or fungi.

The **food pyramid** is a way of expressing the availability of food in an ecosystem at a successive number of trophic levels. The number of producers, always at the base of the pyramid, is high, and the number of consumers at the top of the pyramid is low. The difference in numbers occurs because only a small percentage of the food energy available at one level can be passed on to the next. The total dry weight of food at each level of the pyramid is the **biomass.**

Another phenomenon of an ecosystem is the recycling of minerals. Carbon, nitrogen, and phosphorus typify those minerals that are recycled. Much of the carbon is recycled in respiration, but more is recycled in decomposition, principally by bacteria and fungi.

Nitrogen, which is vital for the synthesis of proteins and nucleic acids, is released to the atmosphere as waste proteins by bacteria. Nitrogen is brought back into the food chain by nitrogen-fixing bacteria that exist on the roots of plants called legumes (peas, beans, alfalfa, and clover). The bacteria trap the nitrogen, form ammonium ions, and make these ions available to plants for amino acid synthesis.

Biosphere

The **biosphere** is the blanket of living things that surrounds the substratum of the earth. The biosphere is composed of living organisms as well as the physical environment. The physical environment includes the rocky material of the earth's crust, the water on or near the earth's surface, and the thin blanket of gas surrounding the earth. All life is confined to a five-mile vertical space around the surface of the earth.

Ecologists study the living components of the biosphere in subunits called **biomes.** A biome is a group of communities dominated by a particular climax community. Deserts, forests, and prairies are examples of biomes. Other examples are the tundra, taiga (the southern edge of the tundra), and temperate forests. Each biome represents a unique situation where the ecosystem is defined by the environment. The broad diversity of living things that characterizes the earth exists in the biome. Each type of living thing is adapted to its own habitat and niche within the biome. The general composition of a biome remains uniform, but local differences arise as a result of population fluctuations, floods, fire, and other ecological factors.

Chapter Checkout

Q&A

1. The two types of Paramecium, *P. aurelia* and *P. caudatum,* are closely related species. Each population, when grown alone under constant conditions, is able to grow and flourish. When put together, *P. aurelia* was unable to survive yet *P. caudatum* continued to flourish. This experiment and results are an example of which of the following?

    a. carrying capacity
    b. density independent
    c. competition exclusion

2. The term that describes two populations living together in a community is _____.

a. mutualism

b. symbiosis

c. commensalisms

3. Organisms that trap the energy in an ecosystem are called _____.

a. consumers

b. predators

c. producers

Answers: 1. c **2.** b **3.** c

CQR REVIEW

Use this CQR Review to practice what you've learned in this book. After you work through the review questions, you're well on your way to achieving your goal of understanding the key concepts of biology.

Chapters 1–4

1. Describe the different components that make up an atom.

2. A _____ is a precise arrangement of atoms of different elements.

   a. compound
   b. protein
   c. carbohydrate
   d. molecule

3. The _____ organelle is the site of protein and lipid processing and packaging.

   a. endoplasmic reticulum
   b. Golgi apparatus
   c. nucleus
   d. mitochondria

Match the following five types of movement with their definitions:

4. diffusion

5. osmosis

6. facilitated diffusion

7. active transport

8. endocytosis

   a. movement across the membrane from a region of lower concentration to a region of higher concentration at the expense of ATP.
   b. the process by which the plasma membrane engulfs particles or tiny vesicles near the cell surface.

   c. movement from high concentration to a region of low concentration that is assisted by proteins.

   d. movement of water across a plasma membrane from a region of higher concentration to lower concentration.

   e. movement of molecules across a plasma membrane from a region of higher concentration to a region of lower concentration.

Chapters 5–6

9. Within the carbon-fixing reactions of photosynthesis, CO becomes attached to a five-carbon compound called _____.

   a. phosphoglycerate
   b. ribose
   c. ribulose diphosphate
   d. ribulose diphosphate carboxylase

10. _____ describes the process by which glucose is broken down to form two molecules of pyruvic acid.

   a. glycolysis
   b. Krebs cycle
   c. chemiosmosis
   d. Calvin cycle

11. The final electron acceptor of respiration is _____.

   a. NAD
   b. O_2
   c. H_2O
   d. FAD

12. For every one molecule of glucose, the Krebs cycle forms _____ molecules of ATP.

   a. 2
   b. 4
   c. 6
   d. 10

Chapters 7, 8, and 30

13. Describe the genetic relevance of Prophase I during meiosis.

14. Explain the fertilization and development process in humans.

Chapters 9–11

15. List the three Mendelian laws of genetics.

16. In the following hypothetical situation, the length of a giraffe's neck is determined at the locus T (tall), where tall necks are dominant over small necks. What are the possible genotypes of the parent generation if the resulting offspring contain a phenotypic ratio of 4–0?

17. What is the genotype and phenotype F_1 generation resulting from a cross between two pink snapdragons?

18. An enzyme called _____ adds new nucleotides to a growing DNA strand whereas _____ creates a permanent bond between nucleotides.

   a. ligase, polymerase
   b. polymerase, ligase
   c. polymerase, helicase
   d. ligase, helicase

19. Describe the lactose operon.

Match the following three types of RNA molecules with their functions:

20. tRNA

21. mRNA

22. rRNA

   a. receives the genetic code in the DNA and carries the code into the cytoplasm
   b. carries the amino acid to the ribosomes for protein synthesis
   c. used to manufacture ribosomes

23. True or False: Hemophilia is a sex-linked condition that is linked to the Y chromosome.

Chapters 12–14, and 31

Match the following four relationships with their definitions:

24. commensalism

25. parasitism

26. synergism

27. mutualism

    a. a relationship where two populations accomplish together what neither population could accomplish on its own.

    b. a relationship where one population benefits while the other population is harmed.

    c. a relationship that is mutually beneficial.

    d. a relationship where one population benefits while the other is neither benefited nor harmed.

28. True or False: Biomes are the blanket of living things that surround the substratum of the earth.

29. True or False: The first primitive cells were called proteinoids.

30. True or False: The endosymbiotic theory suggests that bacteria were engulfed by larger cells and eventually became the mitochondria.

Chapters 15–18

Match the following three classes of Fungi with the organisms they contain:

31. Oomycetes

32. Ascomycetes

33. Zygomycetes

    a. mildews
    b. water molds
    c. bread mold

Match the following five divisions of algae with the types of algae they contain:

34. Pyrophyta

35. Rhodophyta

36. Chlorophyta

37. Phaeophyta

38. Chrysophyta

   a. green algae
   b. kelp
   c. dinoflagellates
   d. diatoms
   e. red algae

Chapters 19–20

39. Describe the structural features of a flowering plant.

40. Describe the four different tissue systems of a vascular plant.

Chapters 21–22

Match the following five classes of vertebrates with the organisms they contain:

41. Agnatha

42. Amphibia

43. Osteichthyes

44. Aves

45. Placodermi

   a. the earliest known fishes with jaws
   b. fishes that have no jaws
   c. frogs
   d. birds
   e. bony fishes

Chapters 23–29

46. The portion of the brain that contains the higher mental activities such as learning and memory is called the _____ .

   a. hindbrain
   b. midbrain
   c. cerebellum
   d. cerebrum

47. The space located at the junction of a neuron and a muscle fiber is called _____ .

   a. a synapse
   b. action potential
   c. resting potential
   d. a neuromuscular junction

48. Vasopressin is produced in the _____ whereas prolactin is produced in the _____ .

   a. posterior pituitary, anterior pituitary
   b. anterior pituitary, thyroid
   c. thyroid, anterior pituitary
   d. posterior pituitary, thyroid

49. Compare and contrast the hydrostatic skeleton, exoskeleton, and endoskeleton.

50. Describe the types of muscle in the human body.

Answers

1. Provide your own answer (see Chapter 2). **2.** d **3.** b **4.** e **5.** d **6.** c **7.** a **8.** b **9.** c **10.** a **11.** b **12.** a **13.** Provide your own answer (see Chapter 8). **14.** Provide your own answer (see Chapter 30). **15.** Mendel's law of dominance; Mendel's law of segregation; Mendel's law of independent assortment (see Chapter 9). **16.** TT × tt; TT × TT; tt × tt **17.** RR:Rr:rr (1:2:1); Red, pink, white (1:2:1) **18.** b **19.** Provide your own answer (see Chapter 10). **20.** b **21.** a **22.** c **23.** False. Hemophilia is linked to the X chromosome. **24.** d **25.** b **26.** a **27.** c **28.** False. The biosphere is the blanket of living things. **29.** False. The first cells were called protocells. **30.** True **31.** b **32.** a **33.** c **34.** c **35.** e **36.** a **37.** b **38.** d **39.** Provide your own answer (see Chapter 19). **40.** Provide your own answer (see Chapter 20). **41.** b **42.** c **43.** e **44.** d **45.** a **46.** d **47.** d **48.** a **49.** Provide your own answer (see Chapter 27). **50.** Provide your own answer (see Chapter 27).

CQR RESOURCE CENTER

CQR Resource Center offers the best resources available in print and online to help you study and review the core concepts of biology. You can find additional resources, plus study tips and tools to help test your knowledge, at www.cliffsnotes.com.

Books

Molecular Cell Biology, 4th edition, by Harvey Lodish *et al* (WH Freeman and Co., 1999) is a commonly used undergraduate biology text that includes topics such as biochemistry, cell biology, molecular biology, and genetics.

Genes VII by Benjamin Lewin (Oxford University Press, Dec. 1999) includes a thorough discussion of genes and their function while incorporating the concepts of molecular biology techniques to explain genetic function.

Biology of Plants, 6th edition by Peter H. Raven *et al* (WH Freeman and Co., 1998) provides a highly illustrated introduction to plant biology.

The Origin of Species: By Means of Natural Selection or the Preservation of Favoured Races in the Struggle for Life by Charles Darwin (Bantam Classic and Loveswept, June 1999) is the renowned publication that examines biological diversity and introduces the concept of natural selection and survival of the fittest.

The Coming Biotech Ages: The Business of Biomaterials by Richard W. Oliver (McGraw-Hill Professional Publishing, November 1999) serves as an informative account of the future of biotechnologies and the social and economic impact of "bioterials" on our society.

Biology: Concepts and Connections, 3rd Edition by Neil A. Campbell *et al* (Prentice Hall College Division, 1999) provides a holistic introduction to biology that includes both print and interactive media supplements that focus on real-world scenarios.

Spatial Ecology by David Tilman and Peter Kareiva (Princeton University Press, Dec. 1997) gives insight to the dynamics and diversity of ecological systems including the predictions of ecological models.

Life: The Natural History of the First Four Billion Years on Earth by Richard Fortey (Vintage Books, 1999) provides an excellent discussion of the evolution of organisms on earth focusing on paleontology and the discovery and identification of fossils.

Hungry Minds also has three Web sites that you can visit to read about all the books we publish:

- www.cliffsnotes.com
- www.dummies.com
- www.hungryminds.com

Internet

Teachers.net at www.teachers.net contains a wide range of biological resources for students and teachers including a library section, review chapters, and amino acid structures.

The Human Genome Project can be followed at www.ncbi.nlm.nih.gov/science96 and contains the map of the human genome as genes are identified and uncovered.

The Franklin Institute Science Museum can be seen at http://sln.fi.edu. The site contains a variety of scientific information including an "ask the scientist" section.

The MAD Scientist at www.madsci.org contains several library references and resources as well as an "ask the scientist" section.

Next time you're on the Internet, don't forget to drop by www.cliffsnotes.com.

Online Magazines and Journals

Visit **Discover Magazine** at www.discover.com to read scientific articles about a wide range of topics of popular interest.

Check out **National Geographic Magazine** at www.nationalgeographic.com to learn about a wide range of topics from earthquakes and volcanoes to sea otters and great white sharks.

The **Science Daily Magazine** at www.sciencedaily.com offers daily scientific headlines of the latest news and provides articles on a wide range of scientific topics.

GLOSSARY

absorption the process in which nutrients enter cells of the villi, then move across the cells and enter blood vessels.

acids compounds that release hydrogen ions (H+) when the compounds are placed in water.

actin a protein filament within the sarcomeres of muscle cells.

action potential occurs when a neuron is displaying a nerve impulse.

active site the portion of an enzyme that interacts with the substrate.

active transport the movement of molecules across a membrane from a region of low concentration to a region of high concentration that requires the expenditure of energy (ATP).

adenosine diphosphate (ADP) a product of adenosine triphosphate (ATP) breakdown.

adenosine triphosphate (ATP) the chemical substance that serves as the currency of energy in cells.

adrenal glands two glands lying atop the kidneys that produce a family of steroids.

aerobic organisms that require oxygen for their metabolism.

algae a large number of photosynthetic organisms that are generally unicellular and not classified as plants.

alleles different forms of the same gene.

alveoli microscopic air sacs that are surrounded by a rich network of blood vessels in mammalian lungs that function in gas exchange; the air sacs are at the end of the bronchioles.

amino acids the building blocks of proteins.

amoeba single-celled organisms with no distinct shape; members of the phylum Sarcodina.

anabolism the process of synthesizing large molecules by joining smaller molecules together.

anaerobic organisms that thrive in an oxygen-free environment.

anaphase a phase during mitosis in which chromatids separate to become visible chromosomes and migrate to opposite poles.

anaphase I a phase during meiosis in which homologous chromosomes separate.

anaphase II a phase during meiosis II in which the centromeres divide and the chromosomes separate from one another.

androgens hormones, such as testosterone, produced from the testes that promote secondary male characteristics.

Animalia the kingdom that includes the animals.

antibodies proteins synthesized by plasma cells that are released into the circulation to the antigen site and destroy the microorganisms by chemically reacting with them.

antibody-mediated immunity the process by which antibodies bind to antigens and destroy the microorganisms in a highly specific manner.

anticodon the complementary codon present on a tRNA molecule.

antigens the immune-stimulating polysaccharides on the surface of cells.

aorta the major artery of the human circulatory system that receives blood from the left ventricle.

appendix a small fingerlike process that may be a vestige of larger organs functional in human ancestors.

archaebacteria ancient bacteria that have a different ribosomal structure, membrane composition, and cell wall composition than modern bacteria.

arteries the channels through which fluid flows away from the heart.

atom the smallest part of an element that can enter into various combinations with atoms of other elements.

atrium a thin-walled receiving chamber in which blood accumulates in fishes.

auditory nerve the nerve within the ear that carries impulses to the brain for interpretation.

autonomic nervous system a subdivision of the peripheral nervous system, which is divided into the sympathetic and parasympathetic nervous systems.

autosomes the 22 pairs of human chromosomes that are not sex chromosomes.

autotrophic certain bacteria that synthesize their own foods.

axon the long extension of a neuron.

bacilli the rod-shaped bacteria (singular, *bacillus*).

bark the structure of vascular plants formed between the phloem and the cork.

bases compounds that attract hydrogen atoms when placed in water.

basophils the white blood cells that function in allergic responses.

bicuspid (mitral) valve the valve that leads into the left ventricle of the human heart.

binomial name the scientific name of an organism, which contains two elements.

biomass the total dry weight of food at each level of the food pyramid.

biome a group of communities dominated by a particular climax community, such as deserts, forests, and prairies.

biosphere the blanket of living things that surrounds the substratum of the earth.

blastocyst a hollow ball of cells resulting after the morula has passed through the Fallopian tubes and enters the female uterus.

blood clotting the process in which platelets adhere to the walls of damaged blood vessels, setting off a series of processes leading to the formation of a patchy mesh at the injury site.

blue-green algae cyanobacteria; members of the kingdom Monera that are photosynthetic and are found in the soil and in freshwater or saltwater environments.

B lymphocytes white blood cells within the lymph nodes; stimulated by microorganisms or other foreign materials in the blood.

Bowman's capsule an enlarged cup-like structure below the nephron in the human kidney.

bronchi two large tubes at the lower end of the trachea (singular, *bronchus*).

bronchioles the branches formed from the bronchi.

capillaries the microscopic blood vessels between the arteries and the veins.

carbohydrates the primary energy source for living things; composed of carbon, hydrogen, and oxygen.

cardiac muscle the involuntary muscle found in the heart; contains actin and myosin filaments.

carnivores animals that eat other animals.

carrying capacity a situation when a population has reached the maximum size that the environment can support.

catabolism the breakdown or digestion of large, complex molecules.

cecum a blind sac that is the meeting point of the small and large intestines.

cerebellum a portion of the hindbrain that serves as a coordinating center for motor activity.

cell body the main portion of the nerve cell.

cell cycle many repetitions of cellular growth and reproduction; divided into interphase and mitosis.

cell-mediated immunity the process in which the T lymphocytes interact with the microorganisms cell-to-cell and destroy them.

cells the fundamental units of living things.

cellular respiration the process by which animals and other organisms obtain the energy available in carbohydrates.

cell wall a strong membrane outside the plasma membrane present in certain cells, such as bacteria and plants.

centriole a cylinder-like organelle that assists in chromosomal migration during mitosis.

centromere the place of attachment of the two homologous chromatids during prophase in mitosis.

cerebrum the portion of the forebrain that controls higher mental activity, such as learning, memory, logic, creativity, and emotion.

cervix the opening at the lower end of the uterus.

chemiosmosis the subdivision of cellular respiration in which the energy given off by electrons is used to pump protons across a membrane and synthesize ATP.

chemoreceptors the specialized receptor cells that transmit smell and taste.

chlorophyll green pigment that makes up a photosystem that absorbs energy from the sun during photosynthesis.

chloroplast an organelle within green plants in which photosynthesis occurs.

chordates animals with rods along their backs, including reptiles, amphibians, birds, and mammals.

chromatid homologous chromosomes joined to each other at the centromere; present during the prophase of mitosis.

chromatin compacted DNA and protein.

chromosomes linear units of DNA.

chyme a soupy liquid formed in the stomach from the churning of the bolus with gastric juices.

circulatory system the transport system in animals.

class a grouping of similar orders.

cocci spherical bacteria (singular, *coccus*).

cochlea a snail-like series of coiled tubes within the skull that assist hearing.

coenzymes organic molecules that act as cofactors, such as NAD and FAD.

cofactors ions or molecules that associate with enzymes and are required for enzymatic reactions to take place.

commensalism a relationship in which one population receives a benefit from an association while the other is neither benefited nor harmed.

community a situation in which populations of organisms each contain a habitat and a niche.

comparative anatomy comparing the anatomical structures of modern day organisms with fossils to yield clues to the type of organisms that roamed earth long ago.

comparative biochemistry the comparison of biochemical processes of modern day organisms with fossils and ancient species; modern biochemistry indicates there is a biochemical similarity in all living things.

compound a collection of molecules.

cone cells cells of the eye that detect color.

consumers the organisms within an ecosystem that meet their energy needs by feeding on the producers.

cork a tough tissue that combines with the phloem to become the bark of vascular plants.

coronary arteries the arteries that supply the heart muscle with blood.

corpus luteum the mass of cells derived from the female follicle that secretes progesterone.

cortex the outer portion of the adrenal gland.

corticosteroids the steroids secreted from the adrenal glands.

cristae the folds of the inner mitochondrial membrane.

crossing over a process during prophase I in which segments of DNA from one chromatid in the tetrad pass to another chromatid in the tetrad.

cyanobacteria see *blue-green algae*.

cytochromes molecules that accept and release electrons in an electron transport system.

cytokinesis the process during mitosis in which the cytoplasm divides into two separate cells; also called cytoplasmic division.

cytoplasm semiliquid substance that composes the foundation of the cell and contains the organelles.

cytoskeleton an organelle within cells consisting of an interconnected system of fibers, threads, and interwoven molecules that give structure to the cell.

cytosol see *cytoplasm*.

decomposers the organisms of decay; usually bacteria and fungi.

dendrites the short extensions of the neuron.

deoxyribonucleic acid see *DNA*.

deoxyribose the five-carbon carbohydrate attached to purine or pyrimidine bases within DNA molecules.

dermal tissue the tissue that functions to protect the plant from injury and water loss and covers the outside of the plant.

diffusion the movement of molecules through a membrane from a region of high concentration to low concentration.

diploid cells having two sets of chromosomes.

diploid nuclei contained within a mass of cytoplasm within cellular slime molds.

disaccharides sugars composed of two molecules.

division see *phylum*.

DNA deoxyribonucleic acid; a double helix nucleotide molecule containing deoxyribose, nitrogenous base, and a phosphate group; contains the genetic information from which amino acids are determined.

DNA fingerprinting a technique that uses electrophoresis to match DNA molecules to one another for identification purposes.

DNA polymerase the enzyme that joins all the nucleotide components to one another to form a long strand of nucleotides.

DNA replication the process by which cells replicate or synthesize their DNA; takes place during S phase of the cell division cycle.

domestic breeding a process of directed evolution that brings about new forms that differ from ancestral stock.

dominant an allele that expresses itself.

ductless glands glands that have no ducts, such as the endocrine glands.

duodenum the first 10 to 12 inches of the small intestine in which most of the chemical digestion takes place.

eardrum the tympanic membrane that receives vibrations from the outer ear.

ecosystems systems formed from the interactions between communities and their physical environments.

ectoderm one of three germ layers that develops into the skin and nervous system.

egg the haploid cell within the female ovary.

elements the fundamental building blocks of matter within all living things.

embryo forms when all the organs of the body have taken shape.

embryology the study of embryonic development.

endergonic reaction chemical reactions in which energy is obtained and trapped from the environment.

endocrine glands glands throughout the animal body that secrete hormones, which help coordinate body systems.

endocytosis the process in which a small patch of plasma membrane encloses particles that are near the cell surface.

endoderm one of three germ layers that develops into the gastrointestinal tract.

endoplasmic reticulum (ER) an organelle comprised of a series of membranes extending throughout the cytoplasm; two types exist, rough and smooth ER.

endoskeleton an internal support system in the echinoderms and most vertebrates that may include a framework of bones and cartilage that serves as a point of attachment for muscle.

endosperm the female tissue that encloses the seed within the angiosperms.

entropy the degree of disorder or randomness of a system.

environmental fitness an individual's ability to adapt to an environment and reproduce.

enzymes proteins that catalyze the chemical reactions within cells.

eosinophils white blood cells whose functions are uncertain.

epididymis the tube in which sperm cells mature.

epiglottis a thin flap of tissue that folds over the opening to the mammalian trachea during swallowing and prevents food from entering the trachea.

epinephrine a hormone produced in the adrenal medulla that increases heart rate, blood pressure, and the blood supply to skeletal muscle.

erythrocytes the red blood cells; disk-shaped cells produced in the bone marrow that have no nucleus; their cytoplasm is filled with hemoglobin to transport oxygen.

erythropoietin a hormone produced by the kidney cells that functions in the production of red blood cells.

esophagus a thick-walled muscular tube located behind the windpipe that extends through the neck and chest to the stomach.

estrogen a hormone produced by the ovaries that stimulates the development of secondary female characteristics.

eubacteria modern bacteria.

eukaryotes cells that contain a nucleus and internal cellular bodies called organelles.

evolution changes that occur within populations and organisms that make individuals able to adapt to their external environment.

exergonic reaction a chemical reaction in which energy is released.

exocrine glands glands, such as the salivary glands, that deliver their enzymes via ducts.

exoskeleton the hard, protective, outer covering of arthropods and mollusks.

facilitated diffusion the movement of molecules across a membrane from a region of high concentration to a region of low concentration that is assisted by proteins.

Fallopian tubes the passageways that egg cells enter after release from the ovaries; also called oviducts.

family similar genera classified together.

fats lipids composed of a glycerol and fatty acids.

fatty acids long chains of carbon atoms with carboxyl groups at one end.

feeding pattern the pattern in which animals obtain their nutrients.

fermentation an anaerobic process in which energy can be released from glucose even though oxygen is not available; occurs in yeast cells.

fertilized egg cell an egg cell that has been fertilized by a sperm cell.

fetus results from a developing embryo at about eight weeks when the embryo is somewhat human looking and the remaining development consists chiefly of growth and maturation.

flavin adenine dinucleotide (FAD) a coenzyme that functions in the production of ATP.

food chain the transfer of food energy from producers to consumers.

food pyramid a way of expressing the availability of food in an ecosystem at a successive number of trophic levels.

food web many interwoven food chains.

forebrain a portion of the brain that consists of the cerebrum, thalamus, hypothalamus, and limbic system.

Fungi a kingdom that includes the yeasts, molds, mildews, and mushrooms.

G_1 phase a phase within interphase of the cell division cycle that prepares cells for DNA replication.

G_2 phase a phase within interphase of the cell division cycle that prepares cells for mitosis.

gametes sex cells of parent organisms; usually haploid cells.

gastrin a hormone produced by digestive glands to influence digestive processes.

gene the functional segment of chromosomes.

gene flow a mechanism of evolution that results when individuals migrate from one group to another and contribute their genes to the gene pool of the new population.

gene linkage the concept of transfer of a linkage group.

gene linkage map a map that pinpoints the location of genes based on their connection to certain marker gene sequences.

gene pool the collection of genes within a population; as changes in the gene pool occur, a population evolves.

genetic drift a mechanism of evolution that occurs when a small group of individuals leaves a population and establishes a new one in a geographically isolated region.

genome the set of all genes that specify an organism's traits.

genotype the gene composition of a living organism.

genus a grouping of similar species (plural, *genera*).

geographic distribution the distribution of species in geographical areas.

geotropism the turning of a plant away from or toward the earth.

gills structures that allow fish to exchange gases with their environment.

glial cells the cells of the nervous system that support, protect, and nourish the neurons.

glomerulus a ball of capillaries that comprises Bowman's capsule in the human kidney.

glottis a slitlike structure at the opening to the mammalian trachea.

glucagon a hormone produced in the pancreas that stimulates the breakdown of glycogen to glucose in the liver.

glucose a carbohydrate with the chemical formula $C_6H_{12}O_6$ that serves as the primary carbon source of living things.

glycogen a polysaccharide composed of thousands of glucose units that serves as the storage form of glucose in the human liver.

glycolysis the subdivision of cellular respiration in which glucose molecules are broken down to form pyruvic acid molecules.

Golgi apparatus an organelle within eukaryotic cells comprised of a series of flattened sacs; the site of protein and lipid processing and packaging; also called Golgi bodies.

Graafian follicle a cluster of cells within the ovary that is derived from egg cells and secretes female hormones called estrogens.

ground tissue the tissue of the vascular plant that is responsible for storing the carbohydrates produced by the plant.

gymnosperms vascular plants having naked seeds, such as the conifers.

haploid cells containing one copy of each chromosome.

hemoglobin a red pigment that binds oxygen and carbon dioxide molecules and carries them through the bloodstream.

herbivores animals that eat plants.

heterotrophic species that acquire food from organic matter.

heterozygous two different alleles that are present for a particular characteristic.

hindbrain the portion of the brain that consists of the medulla, pons, and cerebellum.

histones nuclear proteins that coil DNA molecules.

homeostasis the process in which the internal environment exists at a steady-state equilibrium despite changes in the external environment.

homeothermic animals that can maintain a constant body temperature.

homozygous two identical alleles that are present for a particular characteristic.

hormones biochemical substances produced within plant or animal cells, or glands, that exert a particular effect.

hydrostatic skeleton a water-based skeleton present in many animals (such as the earthworm) that lack structures, such as bone, for muscles to pull against.

hypothalamus the portion of the forebrain that serves as the control center for hunger, thirst, body temperature, and blood pressure.

hypothesis the proposal of a solution to the question within the scientific method.

ileum the final 12 feet of the small intestine.

immune response the stimulation of B and T lymphocytes.

incomplete dominance an allele combination in which two characteristics blend and both alleles can express themselves; one example is red, white, and pink snapdragons.

inner cell mass a group of cells that continues to develop at one end of the blastocyst.

interneuron a type of neuron that connects sensory and motor neurons and carries stimuli in the brain and spinal cord.

interphase the cell division cycle phase in which the cell spends most of its time; includes G_1, S phase (DNA replication), and G_2.

invertebrates the most primitive of the chordates; lack a backbone.

involuntary muscle see *smooth muscle* and *cardiac muscle.*

islets of Langerhans clusters of cells that make up the endocrine portion of the pancreas.

jejunum the second 10 inches of the small intestine.

kinetochore a region of DNA that has remained undivided during prophase of mitosis; binds to the spindle fibers that eventually pull apart the sister chromatids.

kingdom the largest and broadest category of the classification system.

Krebs cycle the subdivision of cellular respiration in which pyruvic acid is broken down and the energy in its molecules is used to form high-energy compounds.

larynx the voicebox of mammals, formed from several folds of cartilage at the upper end of the trachea.

left atrium the chamber of the human heart that receives oxygen-rich blood via the pulmonary vein.

left ventricle the chamber of the human heart in which oxygen-rich blood enters through the bicuspid valve that leads into the aorta.

lens the portion of the eye that focuses the light on the retina.

leukocytes the white blood cells produced in the bone marrow that have various functions in the body, such as immune reaction.

lichens associations between the cyanobacteria and the fungi.

ligaments the tough, fibrous tissues that link bones to one another.

limbic system a collection of structures that ring the edge of the brain and apparently function as centers of emotion.

lipid an organic molecule used to form cellular and organelle membranes, the sheaths surrounding nerve fibers, and certain hormones; includes fats as an energy source.

liver the organ that helps to process the products of human digestion and removes excess glucose from the bloodstream, converting it to a polymer called glycogen for storage.

loop of Henle the segment of the human kidney after the proximal tubule.

lungs the organ where oxygen diffuses into the blood to join with hemoglobin in the red blood cells.

lymph a watery fluid derived from plasma that seeps out of the blood system capillaries and mingles with the cells.

lymph nodes capsule-like bodies that contain cells that filter the lymph and phagocytize foreign particles.

lymphatic system the extension of the circulatory system consisting of capillaries called lymph vessels, a fluid called lymph, and structures called lymph nodes.

lymphatic vessels a series of vessels that return the lymph fluid to the circulatory system.

lymphocytes the white blood cells that are essential components of the immune system.

lysosome an organelle within eukaryotic cells; a droplike sac filled with enzymes used for digestion within the cell.

mammals milk-producing animals.

marsupials the mammals whose embryos develop within the mother's uterus for a short period of time before birth.

medulla the inner portion of the adrenal glands; a swelling at the tip of the hindbrain that serves as the passageway for nerves extending to and from the brain.

meiosis the process by which the chromosome number is halved during gamete formation.

menstruation the process by which the endometrium is released in females.

meristematic tissue the growth tissue; the location of most cell division of vascular plants.

mesoderm one of three germ layers that develops to become the muscles and other internal organs.

metabolism the rapid turnover of chemical materials; involves the release or use of chemical energy.

metaphase the stage during mitosis in which the pairs of chromatids line up on the equatorial plate.

metaphase I the phase during meiosis in which tetrads align on the equatorial plate (as in mitosis).

metaphase II the phase during meiosis II in which the chromatid pairs gather at the center of the cell prior to separation.

midbrain a portion of the brain that lies between the hindbrain and the forebrain that consists of a collection of crossing nerve tracts.

minerals types of nutrients that include phosphorus, sulfur, potassium, magnesium, and zinc.

mitochondrion the organelle that is the site of energy production in eukaryotic cells.

molecule a precise arrangement of atoms of different elements.

Monera the kingdom that includes the bacteria and the cyanobacteria; prokaryotic organisms.

monocytes some of the white blood cells that function in phagocytosis.

monosaccharides sugars that are composed of single molecules.

monotremes the egg-laying mammals that produce milk.

morula a solid mass of cells that develops about six days after fertilization of an egg cell.

motor neuron a type of neuron that transmits impulses from the brain and spinal cord to muscles or glands.

mRNA messenger RNA; the RNA molecules that receive the genetic code in the DNA and carry the code into the cytoplasm where protein synthesis takes place.

multiple alleles a condition in which more than two alleles exist for a characteristic; one example is A, B, AB, and O blood types.

muscle contraction a process in which actin and myosin proteins move within a sarcomere.

mutation a random change in the gene pool of a population that gives rise to new alleles and is the source of variation in a population.

mutualism a living arrangement in which both partners benefit.

myelin sheath a fatty layer of material that covers the axons of nerve cells.

myofibrils microscopic filaments that make up a muscle cell.

myosin a protein microfilament that comprises the sarcomere of muscle cells.

natural selection the concept that random, small variations take place in living things that lead to the gradual development of a species.

nephron the functional and structural unit of the kidney that produces urine and is the primary unit of homeostasis in the human body.

nerve chord also called a spinal cord; a hollow structure that extends the length of the animal just above the notochord.

nerve impulse an electrochemical event that occurs within the neuron.

nerve roots the 31 pairs of projections that extend out along each side of the spinal cord; the sites of axons of the sensory and motor neurons.

nerves bundles of axons bound together.

neuroglia the glial cells together with the extracellular tissue.

neuron a nerve cell.

neurotransmitter a chemical substance that accumulates in the synapse and increases the membrane permeability of the next dendrite.

neutrophils the white blood cells that function in phagocytosis.

nicotinamide adenine dinucleotide (NAD) a coenzyme that functions during respiration to produce ATP.

nicotinamide adenine dinucleotide phosphate (NADP) a coenzyme that functions during photosynthesis to produce ATP.

nitrogenous base the nitrous molecules that make up DNA (and RNA) molecules; two major types are purines and pyrimidines.

nonvascular plants the plants that do not have specialized tissues to transport fluids.

norepinephrine a hormone produced in the adrenal medulla that intensifies the effects of epinephrine.

notochord a flexible rod of tissue extending the length of an animal that provides internal support.

nucleic acids large molecules comprised of nucleotides.

nucleoli the small organelles that make up the nucleus; the site for ribosomal synthesis, assembly, and packaging (singular, *nucleolus*).

nucleotide the unit that makes up nucleic acid; contains a nitrogen base, a phosphate group, and a carbohydrate molecule.

nucleus the organelle within eukaryotic cells that contains the genetic material, DNA.

Okazaki fragments new sections of DNA that are placed along the lagging strand during DNA replication and are joined together by DNA ligase to produce a new DNA strand.

olfactory nerve the nerve that carries the impulse from the nose to the brain for interpretation.

omnivores animals that consume both plants and animals.

oocytes the developed oogonia in a female after the age of puberty.

oogonia primitive egg cells that accumulate in the ovaries before a female is born.

optic nerve the nerve that carries impulses from the eye to the brain.

order a grouping of similar families.

organelles microscopic bodies within the cytoplasm that perform distinct functions.

osmosis the movement of water molecules across a membrane from a region of high concentration to a region of low concentration.

ovary an endocrine gland that secretes estrogens. In plants, the structure of the pistil where the ovules are enclosed.

oviducts see *Fallopian tubes.*

ovulation the process by which an egg cell is released from the follicle and swept into the Fallopian tube where it moves toward to uterus.

ovules the protective structures that contain egg cells produced by the female.

Pacinian corpuscles the touch and pain receptors on the skin, muscles, and tendons.

paleontology the science of locating, cataloging, and interpreting the life forms that existed in past millennia.

pancreas a large, glandular organ lying near the stomach that produces many of the enzymes used to digest food.

parasites organisms that attack living things and cause disease.

parasitism a type of symbiosis in which one population benefits while the other is harmed.

parasympathetic nervous system a subdivision of the autonomic nervous system that returns the body to normal after an emergency.

parathyroid glands glands located on the posterior surfaces of the thyroid gland that produce parathyroid hormone.

pathogenic organisms that cause human disease.

PCR polymerase chain reaction; a technique used to amplify a gene of interest.

peptides small proteins.

peripheral nervous system a collection of nerves that connect the brain and spinal cord to other parts of the body and the external environment.

peristalsis a rhythmic series of muscular contractions that propels the bolus along.

peroxisome cytoplasmic body containing enzymes for digestion.

phagocytes cells that attack and engulf invading microorganisms.

phagocytosis occurs when the vesicle formed from endocytosis contains particulate matter; the process by which cells or microorganisms are engulfed by another cell.

pharynx the cavity at the rear of the mouth that the nasal chambers open into; the throat.

phenotype the expression of genes and the physical characteristics that result.

phloem structures of vascular plants that transport sugars and other nutrients from the leaves to the other parts of the plant.

phosphate group a group derived from a molecule of phosphoric acid that connects the DNA molecules to one another.

phosphate ion a product of adenosine triphosphate (ATP) together with ADP.

photosystem the site within the chloroplast in which sunlight is captured; includes the pigment molecules, proton pumps, enzymes, coenzymes, and cytochromes.

phototropism the bending and turning of the plant stem toward a light source.

phyla related classes grouped together (singular, *phylum*).

physical map a map that locates a gene of interest precisely by showing the actual number of base pairs between genes on a chromosome.

pineal gland a human endocrine gland in the midbrain that regulates mating behaviors and day–night cycles.

pinocytosis when the vesicle formed from endocytosis contains droplets of fluid.

pistil the structure of the flower that contains a stigma, a style, and an ovary.

pith the structure at the center of the stem of vascular plants.

pituitary gland a gland at the base of the brain consisting of the anterior and posterior lobes that secretes several hormones.

placenta the structure that supplies the fetus with nourishment.

placental mammals mammals that have a nutritive connection between the embryo and the mother's uterine wall.

plant hormones hormones that regulate the growth and development of many plants.

plasma a straw-colored liquid composed primarily of water; the fluid portion of blood.

plasma cells large antibody-producing cells derived from B lymphocytes when stimulated.

plasma membrane also called a cell membrane; a membrane composed of lipids, proteins, and phospholipids.

plasmid small circular DNA molecules often used as vectors to transform specific genes into cells.

platelets small disk-shaped blood fragments produced in the bone marrow that serve as the starting material for blood clotting.

polygenic inheritance the condition in which some characteristics are determined by an interaction of genes on several chromosomes or at several places on one chromosome; one example is human skin color.

polymerase chain reaction see *PCR.*

polysaccharides complex carbohydrates formed by linking multiple monosaccharides.

pons the portion of the hindbrain below the medulla and the midbrain that acts as a bridge between various portions of the brain.

population an interbreeding group of individuals of one species occupying a defined geographic area.

predation a relationship in which one population within a community may capture and feed upon another population.

producers organisms within an ecosystem that trap energy (by photosynthesis).

progesterone a hormone produced by the corpus luteum that regulates the buildup of tissue in the endometrium and inhibits the contractions of the uterus.

prokaryotes cells that do not contain a nucleus or internal organelles; include bacteria, cyanobacteria, and archaebacteria.

prophase the first phase of mitosis; involves chromosomal condensation, nuclear membrane breakdown, and the migration of centrioles to opposite poles.

prophase I the first phase of meiotic division, during which crossing over takes place.

prophase II the phase during meiosis II in which the chromatin material condenses and each chromosome contains two chromatids attached by the centromere.

prostaglandins the hormones secreted by various tissue cells that produce their effects on smooth muscles, on various glands, and in reproductive physiology.

proteinoids the primitive polymers formed by the unison of amino acids; able to act as enzymes and catalyze organic reactions.

proteins long chains of amino acid units that are the main molecules from which living things are constructed.

Protista a kingdom that includes protozoa, one-celled algae, and slime molds.

protocells the first cells.

protons positively charged particles within the nucleus of an atom.

pulmonary artery the artery of the human circulatory system that pumps the blood from the right ventricle to the lungs for gas exchange.

pulmonary vein the vein of the human circulatory system that returns oxygen-rich blood from the lungs to the left atrium.

purine a type of nitrogenous base present in DNA molecules containing two fused rings of carbon and nitrogen atoms; two examples in DNA are adenine (A) and guanine (G).

Purkinje fibers the nerves that transfer amplified impulses to regions of the heart to control its function.

pyrimidine a type of nitrogenous base in DNA molecules that has one ring containing carbon and nitrogen atoms; two examples in DNA are cytosine (C) and thymine (T).

recessive the allele overshadowed by the dominant allele.

recombinant DNA DNA molecules that have been altered in some way during the process of genetic engineering or biotechnology.

red blood cells also known as erythrocytes; cells that contain hemoglobin to transport oxygen.

reflex arc the simplest unit of nervous activity; involved in the detection of a stimulus in the environment by sensory nerve endings, followed by impulses that travel via the sensory neurons to the spinal cord.

renal arteries arteries in which blood enters the kidney.

renal veins veins in which blood exits the kidney.

responsiveness the ability of living things to respond to stimuli in the external environment.

resting potential the inactive state of a neuron in which the cytoplasm is negatively charged with respect to the outside of the cell.

restriction enzymes catalyze the opening of a DNA molecule at a "restriction" point; many leave dangling ends of DNA molecules at the point where the DNA has been opened.

retina a single layer containing nerve cells within the eye.

RFLP restriction fragment length polymorphism; a technique using small bits of DNA fragments linked to various diseases.

rhodopsin a light-sensitive pigment of the eye that functions in the detection of light.

ribonucleic acid see *RNA*.

ribosomes organelle bodies that may be bound to the ER that are the sites of protein synthesis in eukaryotic cells; the bodies in which amino acids are bound together to form proteins.

right atrium the chamber of the human heart in which oxygen-poor blood enters through a major vein called the vena cava.

right ventricle the pumping chamber of the human heart from which blood exits.

RNA ribonucleic acid; a nucleic acid produced during transcription that is complementary to a DNA strand; similar to DNA in structure but contains the carbohydrate ribose and the pyrimidine uracil rather than thymine.

RNA polymerase the enzyme that moves along the DNA strand, reads the nucleotides one by one, and synthesizes a complementary mRNA molecule according to the principle of complementary base pairing.

rod cells the cells of the eye that permit vision in dim light.

roots the structures of vascular plants that anchor them to the ground and take in water and minerals from the soil.

rough endoplasmic reticulum ER studded with ribosomes; the site of protein synthesis in eukaryotic cells.

rRNA ribosomal RNA; RNA molecules that function to manufacture ribosomes.

salivary glands the parotid glands, the submaxillary glands, and the sublingual glands that secrete saliva into the mouth.

sarcolemma the muscle cell membrane.

sarcomere the functional unit of the muscle that contains thin actin filaments and thick myosin filaments.

scientific method an orderly process of gaining information about the biological world.

scrotum a pouch outside the male body that contains the testes.

secretin a hormone produced by digestive glands that influences digestive processes.

seedless vascular plants the division Pteridophyta that includes the ferns.

semen a fluid secretion containing sperm and secretions from the prostate gland, seminal vesicles, and Cowper's glands.

semilunar valves two valves found in the pulmonary artery and the aorta.

seminiferous tubules coiled passageways in which sperm production takes place.

sensory neurons neurons that receive stimuli from the external environment.

sensory somatic system a subdivision of the peripheral nervous system that carries impulses from the external environment and the senses.

sepals modified leaves that enclose and protect a growing bud in flowers.

serum plasma from which clotting proteins have been removed.

sex chromosomes one pair among the 23 pairs of human chromosomes; the X and Y chromosomes.

skeletal muscle see *striated muscle.*

slime molds (cellular) amoebalike cells that live independently and unite with other cellular slime molds to form a single, large, flat cell with many nuclei.

slime molds (true) single, flat, very large cells with many nuclei.

small intestine the site of chemical digestion; includes the duodenum, jejunum, and ileum.

smooth endoplasmic reticulum ER with no ribosomes attached.

smooth muscle found in the linings of the blood vessels, along the gastrointestinal tract, in the respiratory tract, and in the urinary bladder; contains few actin and myosin filaments; also called involuntary muscle.

species a group of individuals that share features and are able to interbreed under natural conditions to yield fertile offspring.

spermatogonia primitive cells within the seminiferous tubules that undergo a series of changes and then meiosis to yield sperm cells.

sperm cells haploid cells within the male testes.

S phase the phase within the cell division cycle in which DNA is replicated.

spinal cord the white cord of tissue passing through the bony tunnel made by the vertebrae.

spiracles a series of openings on the body surface of terrestrial arthropods that open into tiny air tubes that assist in gas exchange.

spirilla flexible spiral bacteria (singular, *spirillum*).

spirochetes rigid spiral bacteria.

spleen the site where red blood cells are destroyed; a reserve blood supply for the body.

stamen the structure of a flower that contains a thin, stemlike filament and an anther.

stomata the pores within leaves surrounded by guard cells that regulate the rate of gas exchange, which regulates the rate of photosynthesis. (Singular, *stoma.*)

striated muscle skeletal muscle fiber that appears to be banded due to the presence of overlapping actin and myosin filaments; also called voluntary muscle.

substrate the substance changed or acted on by an enzyme.

survival of the fittest the concept of natural selection that states that the fittest survive and spread their traits through a population.

sutures the immovable joints where bones come together within the skull.

symbiosis the relationship between two populations that live together in a close, permanent, and mutually beneficial association.

sympathetic nervous system a subdivision of the autonomic nervous system that prepares the body for an emergency.

synapse the fluid-filled space separating the end of the axon from the dendrite of the next neuron or from a muscle cell.

synergism a type of relationship in which two populations accomplish together what neither could accomplish on its own.

telophase a phase during mitosis in which the chromosomes arrive at the opposite poles of the cell.

telophase I the phase during meiosis in which the nucleus reorganizes as the chromosomes become chromatin; cytoplasmic division takes place, resulting in two cells.

telophase II the phase during meiosis II in which the chromosomes gather at the poles of the cells and form a mass of chromatin; the nuclear envelope develops, the nucleoli reappear, and the cells undergo cytokinesis.

tendons the connective tissue by which muscles are attached to bones.

testes endocrine glands that secrete androgens; the male reproductive organs located in the scrotum.

thalamus a portion of the forebrain that integrates sensory impulses.

theory a hypothesis that is confirmed through repeated experimentation.

thrombocytes the starting material for blood clotting; also called platelets.

thylakoids membranes that make up the grana in chloroplasts; the actual site of photosynthesis within chloroplasts.

thymosins hormones secreted by the thymus gland that influence the development of the T lymphocytes of the immune system.

thymus gland an endocrine gland in the neck tissues that secretes thymosins.

thyroid gland a gland at the base of the neck that produces several hormones, such as thyroxine and calcitonin.

T lymphocytes white blood cells in the lymph nodes that are stimulated by microorganisms or other foreign material in the blood.

trachea the windpipe of mammals.

tracheae the branching network that extends from holes to all parts of an anthropod body to assist in gas exchange.

tracheids the main conducting vessels of the xylem in most vascular plants.

tracheophytes vascular plants composed of a xylem and phloem.

transcription the process in which a complementary strand of mRNA is synthesized according to the nitrogenous base code of DNA.

transgenic animals animals in which one or more genes have been introduced into the nonreproductive cells.

translation the process by which the genetic code is transferred to an amino acid sequence in a protein.

tricuspid valve a valve that passes blood from the right atrium into the right ventricle.

tRNA transfer RNA; RNA molecules in the cytoplasm of a cell that carry amino acids to the ribosomes for protein synthesis.

trophoblast a layer of cells that forms after fertilization; projections from the trophoblast form vessels, which merge with maternal blood vessels to form the placenta.

tropism the bending or turning response of a plant caused by external stimuli.

turgor pressure the pressure exerted on a plant's guard cells to open.

umbilical cord the source of attachment of the fetus to the maternal blood supply.

urea a component of urine that results from amino acid breakdown in the liver.

ureters tubes that carry waste from the kidneys to the urinary bladder for storage or release.

urethra the path in which urine flows from the bladder to the exterior; the tube within the penis that carries the sperm.

uric acid a component of urine that results from nucleic acid breakdown.

urinary bladder the site where waste products are shipped from the kidney for storage or for release.

urine the product of the kidney; a watery solution of waste products, salts, organic compounds, uric acid, and urea.

uterus a muscular organ in the pelvic cavity of female mammals; also called the womb.

vacuole an organelle found in mature plant cells that stores nutrients and toxic waste.

vagina a muscular organ in female mammals leading from the cervix to the exterior.

vascular bundles arrangements of the xylem and phloem in vascular plants.

vascular plants plants that contain specialized tissues to transport fluids.

vascular plants with protected seeds angiosperms; the most developed and complex vascular plants.

vascular plants with unprotected seeds gymnosperms; vascular plants that contain naked seeds, such as the conifers.

vectors the carriers of DNA genes to be inserted into cells.

veins channels through which fluid flows toward the heart.

vena cava the major vein in the human heart; pumps oxygen-poor blood into the right atrium.

ventricle a pumping chamber for blood to exit from the heart.

vertebrates animals with backbones.

vessels the main conducting vessels of the xylem found in the angiosperms.

virus fragments of nucleic acid surrounded by a protein coat; may attack cells and replicate within the cells, destroying them.

vitamins organic nutrients essential in trace amounts to the health of animals.

voluntary muscle see *striated muscle.*

white blood cells see *leukocytes.*

xylem the structure of vascular plants that conducts water and minerals upward from the roots.

zygote a fertilized egg cell, which is diploid.

Index

Notes

Notes

Notes

Notes

Notes

Notes